Practical Country Living

Improving Your Garden Soil

10 Steps to Healthy Plants and Nutrient-Rich Food

Practical Country Living

Improving Your Garden Soil

10 Steps to Healthy Plants and Nutrient-Rich Food

Michelle Lindsey

Homestead on the
— RANGE —

Abundant Living in Flyover Country

Published by Homestead on the Range

Derby, Kansas

https://homesteadontherange.com/

Homestead on the Range is a Kansas-based small business dedicated to serving country living enthusiasts by supplying them with the innovative resources that they need to succeed. Whether your family's farm or ranch is 5 acres or 500, a business or a hobby, in Kansas or in some other part of the world, our goal is to keep you informed and inspired.

Disclaimer

This publication contains materials designed to assist readers and for educational purposes. While the author and publisher have made every attempt to verify that the information provided in this book is correct and up to date, neither the author nor the publisher assumes any responsibility for any error, inaccuracy, or omission.

ISBN 978-09975261-7-2 (paperback)
ISBN 978-09975261-8-9 (eBook)

Contents

Introduction

Soil is more than just an inert medium. It is a vibrant community that feeds the plants that in turn feed us.

Many gardeners are obliged to start out with less-than-ideal garden soils. Many soils suffer from drainage problems and nutrient imbalances that limit the amount of produce that can be grown on them. Diseases, insect attacks, and poor-tasting vegetables are the result.

The solutions touted to correct these problems are often either complicated or costly. Manufacturers of expensive, hard-to-find amendments all clamor for our attention. Mystique surrounds processes such as composting, convincing us that everything must be done exactly the right way for success. Meanwhile, conflicting information abounds, telling us one day that peat moss will save our garden and the next that it will kill our plants.

The good news? Improving soil and growing vibrant, nutrient-rich produce doesn't have to be this complicated.

About This Book

Improving Your Garden Soil seeks to help you improve the quality of your harvests by cultivating the following benefits:

- Improved plant health.
- Fewer insect pests.
- Increased yield.
- Enhanced produce flavor and texture.
- More nutrient-rich food.

The approach to soil care presented in the following pages has been formed by two decades of gardening experience, mostly with organic row gardening on a problematic, weed-ridden clay soil in a low-lying area. Most recently that experience has been applied to a 100-square-foot raised bed filled with that same clay soil. The techniques used were drawn from years of diverse reading. Writers who have informed the thinking behind this book range from lunatic farmer Joel Salatin to the Kansas State University extension system, from smiling gardener Phil Nauta to square foot gardener Mel Bartholomew, who made soil improvement easier for gardeners by avoiding it altogether. Obviously this represents a wide philosophical spectrum.

The author is happy to report that these methods work.

10 Steps to Soil Improvement

This book simplifies soil improvement by using a 10-step approach:

1. Assessing your current soil.
2. Planning your soil improvement project.
3. Building soil depth.
4. Eliminating toxins, diseases, and pests.
5. Balancing the air and moisture in the soil.
6. Charging up the soil battery.
7. Balancing soil nutrients.
8. Building organic matter and humus.
9. Building a soil community.
10. Maintaining good soil.

This step-by-step process makes it easier to know where to start for beginners. It also helps bring poor soils into production quickly by targeting the most crucial limiting factors first. This will keep soil improvement from becoming overwhelming, and it will help you spend your gardening budget more effectively.

Note that you don't necessarily have to follow each step. If your soil is already in reasonable condition, you probably won't need to build depth or eliminate toxins, for instance. You can start with the assessments and planning process, then skip ahead to, say, building organic matter.

And, once you have achieved satisfactory soil, we'll address how to keep it that way. This book will continue to serve as a reference and a tool for further monitoring and improvement.

Note that this is a practical, hands-on book. Step-by-step directions have been provided to make the learning process easier, but you must do the work and start building experience with your own local soil conditions.

Creating good soil is not hard. It just takes time. But the delicious, tender produce you will harvest as a result makes it worth the effort.

Dig in!

Chapter 1
Assessing Your Current Soil

Before launching into improving your garden soil, it's best to know a little bit about soil in general and your soil in particular. Understanding what soil is, how it forms, and why it matters is foundational to treating it properly. Identifying the strengths and weaknesses of your garden soil in its present state is also important for creating a workable plan that will benefit your garden without breaking the bank.

Keep in mind that as you perform the suggested tests, you will benefit from taking notes. Keeping a garden journal is highly recommended for monitoring the health and progress of your soil and garden in general. However, taking notes on test results on your computer or mobile device is quite sufficient, as well.

What is Soil?

Soil is about half solid material and half void.

The solid portion of soil is made up of two main components:

- Minerals.
- Organic matter.

The largest portion of soil, or about 45% of the total volume, is the mineral component. It forms slowly from the breakdown of the rocks and minerals beneath the surface, also known as the parent material. It is largely the parent material that makes soils distinct from one another. A soil derived from limestone will be different than one derived from sandstone, for instance, as it will contain particles of a different size and nutrient profile.

The organic matter element typically makes up about 5% of soil volume, but it is nevertheless critical because of the role it plays in binding mineral particles together and in providing additional nutrients to plants. Organic matter is derived from decomposing plant and animal matter. In nature, this includes old leaves, grass residue, old roots, fallen tree limbs, animal feces, and dead animals themselves. Soils tend to contain varying amounts of organic matter; those with a high organic matter content tend to be dark in color.

Minerals and organic matter typically cannot be distinguished with the naked eye because they have been broken down by wind, water, temperature changes, and microbial activity.

As for the half of the soil that is void rather than solid material, this space is filled with roughly equal amounts of water and gases. We say *roughly* because the proportion of water and gas changes on a regular basis. After a soaking rain, the water content increases for a time, displacing the soil atmosphere. In dry weather, there is often more gas than water in the soil. Compact soils may have lower percentages of water and gas relative to solid material.

The gaseous part of soil is its atmosphere. Soil has a different atmosphere than any other part of the earth. It includes high levels of carbon dioxide.

The water portion of soil is actually more than water alone. It contains minerals suspended in a solution.

Also present in the soil, although not exactly part of the soil itself, are a large number of organisms, making soil a sort of ecosystem or community. In fact we could think of soil as an apartment complex. The mineral and organic material components of soil are the building itself, while the various organisms are the tenants. These organisms include everything from bacteria to fungi to nematodes to protozoa. But not all soil residents are microscopic—earthworms, ants, millipedes, moles, and others are also community members.

As we have seen, soil is largely created by the decomposition of minerals and organic matter. However, erosion also plays a role. While erosion invariably comes at the cost of the soil in some part of the world, it can also build up the soil in some other part. Receding water often leaves rich soil deposits in its wake, while the wind has dropped fine, deep soils over the Midwest. Some areas have particularly rich soils thought to have been transported by glaciers.

While the soil includes both topsoil and subsoil, in gardening we primarily deal with the topsoil. This is the uppermost soil layer, often present as a distinct boundary between any organic debris on top of the ground and either a fine-grained subsoil or an intermediate layer of coarse, light-colored particles. In some places, human activity has removed the topsoil, either deliberately as when constructing a house or accidentally as seen with some agricultural practices.

Why Soil Matters

All too often, humans are guilty of treating soil like, well, dirt. We walk all over it!

Puns aside, it is true that we tend to severely underestimate soil's importance. It is so much more than a solid surface under our shoes. It is more than an anchor for plant roots (although it is certainly useful for that, too).

To summarize the importance of soil, it sustains life on earth in the following ways:

- Providing a stable surface to support all aboveground activities.
- Storing excess atmospheric carbon.
- Regulating rainfall runoff, thus reducing flooding.
- Storing moisture, thus providing a buffer against drought.
- Filtering heavy metals and other toxins out of groundwater.
- Anchoring plant roots.
- Buffering temperature changes that would be harmful to plant roots.
- Providing nutrients to plants that feed both animals and humans.
- Storing excess nutrients from organic decomposition for later use.
- Offering a safe habitat to countless organisms, both large and small.
- Providing humans with a useful building material (clay).

Inspecting Your Soil

Want to get a handle on how your garden soil is improving? Do-it-yourself test kits never hurt, but there's quite a bit that you can learn from your own five senses (within reason; please don't taste the soil).

Try this. Dig up a cubic foot of garden soil as a sample (that would be a hole one foot long, one foot wide, and one foot deep). Dig slowly, paying attention to how easy or hard the process is. Note whether the soil clings to your shovel or trowel. Keep your eyes open for earthworms and other soil residents as you dig.

Once you have a hole and some soil, grab a handful and look it over carefully. Smell it. Squeeze it, and then crumble it apart again. Notice what objects you dug up besides soil, such as roots, insects, or rocks.

For best results, inspect your soil in spring or fall at a time when it is neither dry nor saturated. Take notes on your findings. You will need this information later on.

Once you have made your observations, read on to find out what they mean.

Texture

Soil texture is determined by particle size:

- *Sand* particles are the coarsest, ranging in size from about 0.05 mm all the way up to 2 mm.
- *Silt* ranks between sand and clay in particle size, the particles ranging from 0.002 to 0.05 mm.
- *Clay* particles are finer than silt particles, measuring less than 0.002 mm in size.

Loam is a soil with a balanced combination of these three particle sizes (about 40% sand, 40% silt, and 20% clay).

Identifying your soil texture is quite simple. Grab a handful of soil and try to squeeze it into a ball. Then see how easy it is to break the ball up again. While you are at it, rub a pinch of soil particles between your thumb and index finger to see how they feel.

If your soil is primarily sand, you will observe the following:

- Visibility of individual soil particles.
- Irregular particle shape.
- A gritty texture.
- Light feeling on the end of a shovel.
- No tendency toward compaction (you will not be able to shape it into a ball).

If your soil is primarily silt, you will notice:

- No clumps.
- A silky-smooth texture, like flour (never sticky).
- Light feeling on the end of a shovel.
- Moderate ability to be compacted (can be shaped into a ball, but the ball remains pliable and may fall apart altogether if poked).

A primarily clay soil will have the following characteristics:

- Shiny appearance when wet.
- Sticky feel when wet and dusty feel when dry.
- Heavy feeling on the end of a shovel.
- Ease of compaction (readily forms a ball, but the ball is difficult to break apart again, especially after it dries).
- Tendency to expand when wet and bake into cracks when dry.

Loamy soil has its own unique characteristics:

- Soft feel.
- Similarity to chocolate cake crumbs in texture and appearance.

(If you want a more precise approach to identifying your soil type, keep reading. It's coming up later in this chapter.)

Each soil type has its own advantages and disadvantages.

Because sand particles are so large, there is a lot of space between them, which allows water to pass through easily. Sandy soils, therefore, drain well—sometimes a little too well. Plants growing in sand tend to die off during a drought unless watered very frequently. Sand also retains nutrients poorly.

One of the advantages of sand, however, is that it allows plant roots to breathe and grow easily. It is also easy to work.

Silt tends to be fertile and generally makes a better soil for most garden and field crops than either sand or clay. However, because the particles in silt are smaller than those in sand, they are more prone to compaction. When the particles are pressed tightly together, water cannot drain from the soil easily and may drown plant roots. On the other hand, when not compacted, silt tends to erode quite easily. Silty soils warm up slowly in the spring.

But this is not to say that silty soils are all bad. Silt has excellent nutrient retention abilities. As an added bonus, silt is extremely easy to work as long as it is not waterlogged.

Since clay compacts so easily, water has a difficult time passing through it and draining from the soil. This can result in a waterlogged garden after a heavy rain, which in turn means reduced levels of soil atmosphere. Clay soils are particularly prone to forming impenetrable layers of hardpan because their particles carry strong negative charges, thus inviting bonding with other substances. When they dry out, clayey soils can be next to impossible to work. But don't even think about trying to compensate by working the soil when it is wet—wet clay is guaranteed to form heavy clods when tilled. Clay soil also warms slowly in the spring.

The good news about clay is that it retains dissolved nutrients extremely well, and it can hold moisture during dry weather.

Loam generally has all the strengths of its constituent parts without their weaknesses. It holds water better than sand, but drains better than silt and clay, all the while keeping an ideal level of air between its particles. Loam also tends to be fertile because it retains nutrients well. Loam is easy to work, and it requires little maintenance.

Obviously, then, loam is the easiest soil to care for, but don't worry if your spadeful of dirt turns out to be one of the other three. For one thing, some plants, particularly those used for landscaping or native plant gardens, prefer a little more sand or a little more clay. But any soil can be improved with organic matter, as we will see later on.

Structure

Soil texture and soil structure are not the same. Texture refers to the size of the individual soil particles, while structure refers to how those particles are aggregated. Soil structure is partly determined by texture, partly by ionic charge, and partly by a variety of disturbances:

- Tillage.
- Trampling.
- Fertilization.
- Root growth.
- Wet/dry cycles.
- Freeze/thaw cycles.
- Activity of fungi, microbes, earthworms, and other soil life.

While structure is harder to identify than texture (scientists usually check it out with a microscope), it is actually of greater practical importance to gardeners because it influences the ability of air, water, earthworms, and roots to move through the soil. Fortunately, a precise classification is not necessary. Just return to the cubic-foot hole you dug and do the following:

1. Use a sharp trowel or spade to take a slice from the side of the hole.
2. Being very careful to keep the slice intact, lay it out on a flat surface for examination.
3. Note the size and uniformity of the particles.

There are many different terms used to describe soil structure, but you should be able to roughly classify your soil along these lines:

- **Blocky.** Peds (aggregates) consist of six or more flat or slightly rounded surfaces.
- **Columnar.** Similar to prismatic peds, but with distinct, rounded tops.

- **Crumbly.** Soft. Small, porous aggregates that are weakly held together. Good garden soil.
- **Granular.** Spherical or very irregular peds. Small, nonporous aggregates that are strongly held together. Ideal garden soil.
- **Lenticular.** Peds thick in the middle, thinning toward the edges. Typically forms overlapping horizontal layers.
- **Massive.** No obvious structure. All particles are the same size and cling together in large, uniform masses.
- **Platy.** Breaks into flat, plate-like layers or crusts. Typically forms overlapping horizontal layers. Poor permeability.
- **Prismatic.** Peds are bounded by flat to rounded vertical faces with flat, somewhat indistinct tops. Distinctly longer vertically.
- **Single-grain.** No obvious structure. Particles do not form aggregates.

Most soils will benefit from the addition of organic matter to loosen up problematic aggregates.

Also, avoid working the soil when wet, as this tends to destroy the structure.

Color

What color is your garden soil? Does it make a difference?

Yes! Some factors that alter soil color include:

- Aeration.
- Saturation.
- Mineral content.
- Degree of weathering.
- Organic matter and humus content.

All of these considerations are important to the health of your garden plants.

Dark brown to black colors are associated with rich soils, and with good reason. Rich dark brown or black soils typically contain ample humus and organic matter. The darker the color, the more decomposed the organic matter is—in other words, a greater percentage of organic matter has broken down into humus.

Also, very dark soils generally contain sodium, as sodium causes organic matter and humus to disperse more evenly throughout the soil.

But truly black soils can also indicate complete saturation and high levels of anaerobic bacteria. Fortunately, there is no mistaking this type of black with the rich black of healthy soils. Unhealthy black soils look and smell disgusting—they are basically anaerobic rot vats.

Reddish and red-brown soils typically indicate high levels of hematite, an iron oxide. The depth of the red is a good indicator of the efficiency with which the soil drains. Bright reds indicate excellent drainage and infrequent saturation. Pale reds suggest that nutrients have leached from the soil due to a lack of organic matter.

Orange soils are seen when iron has been precipitated out of water. Thus, an orange color suggests that the soil is periodically waterlogged. Or if your soil was previously some other color and has suddenly turned orange for the first time, it could indicate that the problem is excess iron in your water supply.

Yellow soils indicate the presence of goethite, another iron oxide mineral. As with red soils, the brighter the color the better the drainage. As also mentioned with reds, pale yellow soils have been leached and would benefit from some additional organic matter to retain nutrients.

Note that intermediate tan colors are formed when organic matter is added to soils containing iron oxide.

What happens if a red or yellow soil is submerged for an extended period of time? The aerobic bacteria normally present in the soil eventually run out of oxygen and go dormant, leaving the field open for anaerobic bacteria. Anaerobic bacteria convert ferric iron to ferrous iron in their metabolic processes. Ferrous iron has no color; thus as the anaerobic bacteria feast and multiply, the color disappears from the soil. Meanwhile, with water displacing the oxygen in the soil, the iron compounds that form the mineral element of the soil stop oxidizing and producing rust colors, resulting in a more gray appearance.

Grayish soils, therefore, are found in areas with high levels of both moisture and iron. If a gray soil has a bluish or greenish cast, sulfur may be present. A mottled gray soil, as opposed to a uniform gray or blue-gray soil, suggests that the soil is waterlogged at times and fairly dry at other times.

But in any case, gray soils are definitely a cause for concern, as they indicate frequent saturation and poor drainage.

A white layer between the topsoil and the subsoil usually indicates a high percentage of sand. Leaching causes the color (or lack thereof). Plenty of organic matter will be needed to correct this situation.

White coloring in the subsoil is typically due to the presence of calcium carbonate. However, white soil containing calcium carbonate may be seen at the surface of the ground if exposed by erosion. This type of soil typically occurs only in desert areas, as regular rains prevent calcium from accumulating.

If you have a healthy dark brown or black soil, great! However, a bright red, yellow, or tan soil will work, too, with some improvement.

On the other hand, pale and gray soils are more problematic, signaling serious drainage problems that will have to be overcome. Likewise, white soils will require extensive improvement. If you have poor soil, don't worry; the following chapter provides alternatives you may want to consider.

Table 1 Soil Colors

Color	Meaning
Dark brown to rich black	Rich soil, ample organic matter and humus, adequate sodium levels
Black (usually with foul odor)	Complete saturation, high levels of anaerobic bacteria
Bright red to red-brown	High levels of iron, excellent drainage
Pale red	High levels of iron, low nutrient content, insufficient organic matter
Orange	High levels of iron, periodic saturation; sudden color change could be caused by excess iron in water supply
Bright yellow	High levels of iron, good drainage
Pale yellow	High levels of iron, low nutrient content, insufficient organic matter
Tan	High levels of iron, good levels of organic matter
Gray	High levels of iron, prolonged saturation, high levels of anaerobic bacteria
Mottled gray	High levels of iron, sporadic saturation
Blue-gray to green-gray	High levels of iron, high levels of sulfur, prolonged saturation, high levels of anaerobic bacteria
White layer between topsoil and subsoil	Sandy soil, low nutrient content, insufficient organic matter
White subsoil	Insufficient soil moisture, erosion if white subsoil is visible on surface

Smell

The smell of your soil can tell you a great deal.

A rich, sweet, earthy smell is a good sign. It means that your soil microbe community is healthy and hard at work. That smell is produced by fungi and actinomycetes, bacteria that somewhat resemble fungi. Their work in decomposing organic matter produces gases that give healthy soil its distinctive aroma. These garden good guys are also responsible for the fresh smell you notice when it rains after a long dry period.

Most unpleasant soil odors can be attributed to anaerobic conditions, usually caused by poor drainage. These smells might include the following:

- Ammonia-like.
- Sulfurous, like rotten eggs.
- Just plain foul rot smells.

A sour smell may indeed indicate a sour soil, that is, one that has become excessively acidic. Some pH correction may be in order, depending on what you want to grow.

Some soils don't really smell bad; they just don't smell like much of anything. If your soil is fragrance-free or has a bland mineral smell, it is probably lacking in biology. Never fear! Your friendly soil community can be rebuilt over time.

Contents

As you dig up a cubic foot of soil, keep your eyes open for other objects that can affect your gardening experience.

How many earthworms do you see in a cubic-foot sample? Healthy garden soil should have at least 10 worms per cubic foot. Earthworms will avoid very dry and very wet soils, as well as soils that contain harmful chemicals. They also require organic matter to eat. Therefore, a worm-free soil suggests a problem with toxicity, drainage, lack of moisture, or lack of organic matter. A small worm population indicates the same problems, but with less severity.

Then there are the other multitudinous creepy-crawlies you may see. Many are beneficial, as they consume pests and create burrows and tunnels that permit air to flow through the soil. But some insects, such as grubs and cutworms, are pests that will eat your plants. Identifying these bugs may be difficult when they are in the larval stage, as they likely will be if you are finding them in your soil. The good news is that improving your soil will keep them in check without indiscriminately harming the good guys.

You may also see fungi. White, threadlike filaments are mycelia, the vegetative parts of a fungus colony. Mycelia indicate a soil high in organic matter. Beneficial organisms are at work breaking that organic matter down so that your plants can absorb the nutrients.

One thing you should look for is plant roots, whether they are from weeds or from garden plants. Extensive, spreading root systems are a good sign because they indicate that the soil is loose enough to be conducive to root health. Stunted roots often indicate soil compaction.

Do you see organic matter and residue in the soil? Ideally, you'll see both undecayed plant residue and decomposed organic matter (dark, loose soil).

Rocks could be an issue, depending on how many there are. Root crops such as carrots will become deformed trying to grow in rocky soil.

Reading the Plants

Did you know you can learn a lot about your garden soil by simply looking at the plants it supports? Leaves, stems, flowers, and fruits provide clues about the soil moisture, pH, and nutrient levels.

If you already have a garden, take a look at your garden, orchard, and landscaping plants to see what they are telling you. If you don't have a garden, take a look at the weeds growing on your prospective garden site. The prevalent weed species can divulge necessary information.

What Your Garden Plants Tell You

First, let's start with what a healthy plant looks like. A plant that is growing in good soil has a vibrant appearance. Its leaves are glossy due to the lipids it has produced from abundant nutrient intake, and its produce has a sweet, complex flavor for the same reason (when in doubt, ask a young child to taste-test your produce; small children are generally attuned to good flavor, especially when it comes to vegetables).

On to the distress signals.

Table 2 Leaf Signals

Signal	Meaning
Brown spots	Manganese toxicity
Yellowing of leaf margins and between veins of young leaves	Soil saturation, or iron or manganese deficiency
Yellowing of leaf margins and between veins of all leaves	Potassium, magnesium, or zinc deficiency
Uniform yellowing of young leaves	Sulfur or zinc deficiency
Uniform yellowing of old leaves	Nitrogen or molybdenum deficiency
Uniform yellowing of all leaves	Can indicate many problems, including insufficient soil moisture, soil saturation, manganese toxicity, or mercury pollution
Blackening between veins	Boron toxicity
Scorching of leaf margins	Potassium deficiency
Scorching of whole leaf	Usually insufficient soil moisture; phosphorus deficiency is also possible
Cupping of old leaves	Calcium deficiency
Cupping of all leaves	Herbicide drift or magnesium deficiency
Twisting, distortion	Herbicide drift, walnut toxicity, or molybdenum deficiency
Tiny leaves	Nitrogen deficiency

Tiny leaves with puckered margins	Zinc deficiency
Wilting	Insufficient soil moisture or soil saturation
Loss of lower leaves	Phosphorus deficiency
Loss of all leaves	Soil saturation

Table 3 Stem Signals

Signal	Meaning
Tip death	Insufficient soil moisture, or copper deficiency
Rot at soil line	Soil saturation
Slow but normal growth	Soil compaction, or zinc or manganese toxicity
Slow, distorted or swollen growth	Herbicide residue in soil
Stunting, followed by wilt and sudden death	Potassium or copper deficiency, or walnut toxicity

Table 4 Fruit and Seed Signals

Signal	Meaning
Blossom end rot	Calcium deficiency
Delayed fruit set	Herbicide drift
Reduced yields	Herbicide drift or arsenic pollution
No fruit	Nitrogen toxicity
Reduced seed production	Magnesium deficiency
Failure to set seed	Boron deficiency
Chronic failure to germinate	Soil compaction or heavy metal pollution

Table 5 Root Signals

Signal	Meaning
Stunting, deformity	Copper toxicity
Club root	Calcium deficiency
Roots close to the surface of the soil	Soil saturation, compaction, or erosion
Dark, limp roots	Soil saturation
Potato scab	High pH

What Weeds Mean

The prevalent weeds in an established garden or untilled plot can offer clues about the growing conditions at play. Granted, some weeds may have originated from outside sources, perhaps traveling to the site via lawn seed, mulch, or muddy boots. However, for weeds to be become well established, the environmental conditions, including soil conditions, must be to their liking. For this reason, weed identification can offer clues about soil type, pH, drainage, and much more. Just keep in mind that a lone weed may not mean anything; you are looking for a preponderance of one or two species in a specific location.

A comprehensive encyclopedia of weeds and their meanings is far beyond the scope of this book. You will want to find a resource that can help you identify weeds and find out what their preferred growing conditions are. Your local extension service may offer relevant resources. You may also be able to pick up a regional field guide from your library or favorite bookseller.

The following are common garden weeds and what their presence indicates about the soil:

- **Amaranth.** High nitrogen.
- **Annual bluegrass.** Wet soil, compaction, low fertility.
- **Bermuda grass.** Clay.
- **Bindweed.** Clay, saturation, compaction.

- **Broom sedge.** Low calcium, insufficient humus.
- **Burdock.** Compaction, moist soil, high potassium, low calcium.
- **Buttercup.** Saturation, low pH.
- **Cattail.** Saturation.
- **Chickweed.** Saturation, compaction, high pH, high nitrogen.
- **Chicory.** Compaction, high pH, high nitrogen.
- **Clovers.** Compaction, low nitrogen, high potassium.
- **Common groundsel.** Fertility.
- **Cornflower.** Sand.
- **Crabgrass.** Compaction, low fertility.
- **Daisy.** Dry soil, low fertility.
- **Dandelion.** Low pH, compaction, disturbance, high potassium.
- **Dock.** Saturation, low pH.
- **Foxtail barley.** Saturation, compaction, low calcium, high magnesium.
- **Goldenrod.** Wet soil.
- **Goosefoot.** High pH.
- **Henbit.** High nitrogen.
- **Horse nettle.** Hardpan.
- **Horsetail.** Sand, saturation, low pH, low calcium.
- **Joe-pye weed.** Saturation.
- **Knapweed.** High potassium, low phosphorus, low calcium, insufficient humus.
- **Knotweed.** Clay, compaction.
- **Lambsquarters.** High nitrogen, low phosphorus, high potassium.
- **Morning glory.** Hardpan.
- **Mosses.** Saturation, low pH, low fertility.
- **Mullein.** Sand, dry soil, low pH, low fertility.
- **Nutsedges.** Wet soil.
- **Oxeye daisy.** Saturation, low pH, low phosphorus, high potassium, high magnesium.
- **Plantain.** Clay, low pH, compaction, disturbance.
- **Pokeweed.** Fertility.

- **Prostrate spurge.** Compaction.
- **Purslane.** Sand, disturbance, high phosphorus.
- **Quackgrass.** Clay, saturation, compaction, iron/manganese imbalance.
- **Queen Anne's lace.** Dry soil, high pH.
- **Ragweed.** Low fertility, disturbance.
- **Sandbur.** Sand.
- **Sheep sorrel.** Sand, dry soil, low pH, low nitrogen, low calcium.
- **Spotted spurge.** Dry soil, compaction.
- **Stinging nettle.** Low pH, low calcium.
- **Thistles.** Compaction, dry soils.
- **True chamomile.** High pH.
- **Vetch.** Low nitrogen.
- **Violet.** Saturation.
- **Wild garlic.** Clay, compaction.
- **Wood sorrel.** Clay, low pH, high magnesium, low calcium.
- **Yarrow.** Dry soil, low potassium.

Home Soil Testing

Besides a general inspection of soil texture, structure, color, smell, and contents, there are many tests you can perform on your soil to find more specific information. Fortunately, most are quite simple to carry out at home. Some may require special kits or equipment, but for others you will need nothing more than a trowel.

Soil Type

While you may already have an idea of your soil type, here is a way to evaluate it more precisely:

1. Find a one-quart mason jar with a tight-fitting lid.
2. Use a trowel to dig a soil sample to a depth of about eight inches.

3. Place the sample in a large bowl.
4. Collect several samples from different parts of your garden and place them in the bowl.
5. Mix the samples together, picking out any obvious rocks, debris, and pieces of organic matter as you work.
6. Fill your mason jar half full with soil from the bowl.
7. Add enough clean tap water to fill the jar to three quarters full.
8. Add a teaspoon of liquid dish soap (this will help your soil particles separate into layers).
9. Tighten the lid and shake the jar for at least three minutes.
10. Put the jar in a location where it will not be disturbed for about 48 hours.

You are waiting for the soil to settle out into distinct layers. Gravel will sink to the bottom, with sand above that, silt above that, and clay above that. If the soil refuses to settle out and instead remains in the water in a cloudy mixture, it is clay. White grit at the bottom of the jar accompanied by grayish clouding of the water is chalk. Particles floating on top of the water are organic matter.

Look at the sand, silt, and clay portion, noting the proportion each layer occupies and estimating a percentage. You can use a ruler for greater precision, but this is not strictly necessary.

Once you have determined the percentages of sand, silt, and clay in your soil, look at the following soil triangle to identify your specific soil type:

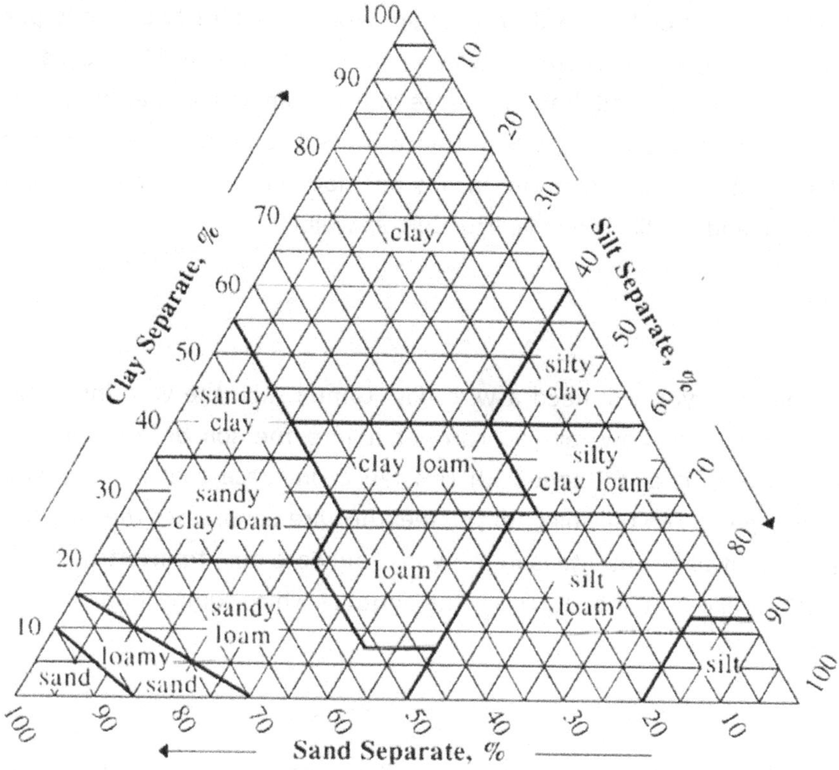

Soil triangle courtesy of USDA NRCS

Percolation

The percolation test determines how well your soil drains. It is quite simple:

1. Return to your previously dug cubic-foot hole, or dig a new one.
2. Fill the hole to the top with water, and walk away for the day. This gets the soil wet for a more accurate reading.
3. The next day, fill the hole up to the top with water again.
4. Measure the water level every hour until the hole is empty, noting how many inches the water drops per hour.

Ideally, the water level will drop at an average rate of two inches per hour, although anywhere between one inch and three inches is acceptable. If your soil drains at a rate of less than an inch per hour, you have a serious drainage problem that will need to be resolved. Soil that drains at a rate of over three inches per hour does not retain enough moisture and will likely need correction, as well.

Compaction

For this test, you will need a wire flag. Gently slide the wire into your garden soil. See how far you can get it into the soil before you feel resistance. If it does meet with resistance, increase the pressure you apply to get it into the soil until you feel the wire bend. Once the wire has bent, pull it back out. Measure how far you were able to insert the wire into the soil before it bent. This is the depth at which compaction starts.

pH

There are several ways to test soil pH. One simple test is to pour vinegar on a sample and see if it fizzes. That indicates alkaline soil.

For more accurate results, however, you probably want to invest in a more sophisticated testing method, especially since home pH testing is very inexpensive.

Probably the easiest method of testing pH is using a test kit. These kits usually include a container to put your sample into, a capsule containing a powder, a color-coded chart for evaluating results, and instructions on how to perform the test. As long as your testing powder is fresh (preferably no more than a year old), you should get satisfactory results. As an added bonus, a pH test kit like this will often include tests for NPK (nitrogen, phosphorus, and potassium), as well.

For the more technologically inclined, there are many pH meters on the market, some good, some not so good. The better models are quite precise. Meters are effective as long as the batteries are new and you are careful to follow the instructions for performing the test and caring for the meter afterward.

NPK

NPK is easy to test at home. You will need a kit, but these are usually quite inexpensive. NPK kits include the same elements as the pH kits—a container, a powder, a chart, and instructions. In fact, your best bet is usually to buy both the NPK and the pH kits packaged together. Keep in mind that this type of soil test works best when the powder is fresh (no more than a year old).

While an at-home NPK kit usually will not tell you the precise amounts of NPK in your soil, it does give you a general idea of whether the levels are adequate for plant needs or in some degree of excess or deficiency. If you follow a holistic, natural approach to soil improvement, this general information will be quite sufficient for home gardening purposes.

Advanced Soil Tests

Not all soil testing can be done at home. Specialized information must be obtained from a lab. Other types of testing can be carried out by the gardener, but require expensive equipment. For this reason, advanced soil testing and its associated costs are not for every home gardener. But for a market gardener or even a home gardener struggling with a really knotty soil problem (or perhaps just the obsessed gardener), advanced soil testing may be in order.

Lab Tests

Your local extension service probably offers garden soil testing. For those who need information beyond the basic pH and NPK levels, this is highly recommended. A good lab test can offer a more precise picture of the nutrient levels in your garden soil; a really good lab test will check for those often-overlooked micronutrients, as well. Some labs may test for other soil constituents of importance, such as organic matter, while others may check for the presence of heavy metals.

When choosing a lab test for your garden, check carefully before you spend any money to make sure that it will provide the information that you need. Again, you can test for pH and NPK cheaply at home, so there is no need to pay extra for this service.

Also, don't waste your money by conducting the test improperly. The soil testing service should provide detailed instructions for correctly preparing the soil sample—follow them to the letter!

Many labs will follow up with recommendations. However, they are typically focused on the use of chemical fertilizers and therefore not useful for organic home gardeners.

Albrecht Test

Think of this as super-advanced lab testing. It is named for Dr. William A. Albrecht, an agronomist and microbiologist who started investigating soil in the 1920s. The Albrecht test uses acids to extract nutrients from the soil sample. It provides a fairly accurate picture of the total nutritional content of the soil. It also tests determines base saturation and cation exchange capacity (CEC):

- *CEC* is the measure of the total negative surface charges in the soil. It is important because it is a measure of how many nutrients your soil can hold on to. An ideal CEC is between 25 and 40. Soil with a CEC lower than the recommended range is probably too sandy with not enough organic matter, while a higher CEC indicates too much clay or organic matter.
- *Base saturation* is the percentage of soil exchange sites that are occupied by the positively charged ions Ca^2+, Mg^2+, K+, and Na+.

Ideal soils typically test 60% to 75% calcium, 7% to 15% magnesium, 2% to 5% potassium, 0.5% to 3% sodium, leaving 10% to 15% of the exchange sites to be occupied by hydrogen.

Because the Albrecht test focuses on the total nutrients in the soil rather than those that are readily available to plant roots, it is best conducted on a regular basis to monitor long-term trends.

You will have to spend some time looking for a lab that does Albrecht testing. Also, keep in mind that most of the labs offering these kinds of tests are geared toward organic agriculture on a commercial scale, and can be very expensive. Again, don't waste your money—prepare and ship the soil sample exactly as instructed.

Reams Test

The Reams test was named for its creator, Dr. Carey Reams, but is sometimes referred to as the LaMotte test. Like the Albrecht test, the Reams test uses acid extraction methods to determine soil nutrient levels. However, the acids used are weaker, having been specifically designed to mimic the natural chemicals given off by plant roots. Thus, the Reams test presents a picture of what nutrients are readily available to plants, catching those baffling problems such as plants grown on a limestone-based soil and yet showing symptoms of calcium deficiency.

A good lab will tell you how your soil compares to the ideal when sending you back the results, and may even make specific recommendations about how to amend the soil. Keep in mind that the following ratios are desired:

- 10:1 calcium to magnesium (7:1 for very sandy soil).
- 2:1 phosphate to potash (4:1 for very sandy soil).
- 1:1 potash to sulfate.

Once again, it will take some time to find a lab that does Reams testing, and the testing can be expensive. However, the test is more useful than the Albrecht test for immediate application.

Brix

Here's a form of testing for the truly obsessed—brix.

Simply put, brix is a measure of the sugar content of a plant or other substance.

To be more specific, Brix expresses the weight of dissolved sugar as a percentage of the weight of the entire solution. Pure water has a brix of 0%. A solution of 5 grams of sugar to 95 grams of water would have a brix of 5%.

Advocates of high-brix gardening say that plants with high brix levels are vibrant and healthy—and vibrant, healthy plants resist insect pests and diseases. They can also stand up to a light frost a little better.

High-brix fruits and vegetables generally have a sweeter, more appetizing flavor than their low-brix counterparts. However, brix proponents claim that there are other benefits to growing and eating high-brix foods. High-brix plants have a superior aroma and may be more digestible. They may even have a higher nutritional content (with some caveats, as we shall see).

Brix can be used to positively determine if a fruit is ripe, as the sugar content of a fruit increases dramatically while ripening.

Finally, high-brix produce keeps surprisingly well.

Brix testing is not a method of soil testing per se. It actually involves testing the juices of garden produce. However, the brix of the produce is directly related to soil health.

Proponents of high-brix gardening agree that soil microbe health is directly correlated with high brix levels in produce. Therefore, while chemical fertilizers can provide brief boosts in brix, they cannot maintain high levels over long periods of time unless special high-sugar hybrid plants are used. Organic fertilizers applied directly to plant leaves can also give a temporary increase in brix. In the long run, however, growers of heirloom and non-hybrid plants must focus on feeding the soil.

Brix is measured with a refractometer. As its name suggests, a refractometer operates on the principle of refraction. When a light beam passes through a liquid solution, it bends, or refracts. The more solids (e.g., sugars) that are suspended in the solution, the more the light refracts. A refractometer is simply a device used to measure refraction.

Two types of refractometers are commonly sold:

- Analog (optical).
- Digital.

An analog refractometer uses a prism and an external light source to operate. A few drops of solution are placed on the prism, the refractometer is held toward a light, and the results are read on a scale.

A digital refractometer works on the same principle, but shines its own light on the prism from an LED. A sensor takes the measurement, a computer calculates the results, and a screen displays the brix reading.

Note that refractometers specifically sold for testing honey are made to display the inverse of a brix reading—that is, they display the moisture content of the sample and not the percentage of dissolved solids. For this reason, you will not want a honey refractometer, but something compatible with gardening purposes.

The reading displayed on the refractometer is not a true measurement of the sugar dissolved in the substance being tested. Any dissolved solids can cause light to refract; therefore all dissolved solids are included in the refractometer results. This includes minerals, proteins, and carbohydrates other than sugar, as well as other non-nutrient solids. Adding fertilizer to a glass of water, for instance, will change the brix reading according to the refractometer.

Of course, this may still help you reach your goal in measuring brix in the first place—increasing the overall nutrient content of your produce. The more minerals, the better.

However, the main disadvantage of using the refractometer is that it cannot tell us exactly what is changing the reading. The only way to know for sure what sugars, minerals, and other solids are in our food is to do a complete nutrient analysis—a little too costly for everyday kitchen use.

Other factors that may influence refractometer results include:

- Sample preparation.
- Sample settling.
- Plant part tested.
- Ambient temperature.

Digital refractometers are programmed to compensate for some of these variables (particularly temperature). To ensure the most accurate results, however, try to test the same part of the plant at the same time of day in every test. Testing in similar weather conditions is also preferable.

So what is an ideal brix level? Most plants show markedly improved vitality and pest resistance when their brix levels hit 12 degrees on a refractometer, although with care many can go far higher than that.

The PC Meter

The PC meter is named for Dr. Phil Callahan, a scientist who researched the paramagnetic (weakly magnetic) properties of rocks and soils all over the world. He found that the most productive soils tended to be those with a stronger paramagnetic field.

A PC meter is used to determine the paramagnetism of soil. Unfortunately, the PC meter is rather expensive, so gardeners who don't want to spend the money are typically encouraged to assume the paramagnetism of their soils is too low.

The PC meter measures paramagnetism in centimeter/grams/seconds (CGS). The minimum acceptable level is 100 CGS. Excellent soil tests at over 700 CGS, and perhaps even over a thousand.

Chapter 2
Planning Your Soil Improvement Project

Now that you have gathered useful information on the current quality of your garden soil, you are ready to use that information to make a plan.

The testing and planning steps of soil improvement are important because gardening can quickly become an expensive hobby. Proper soil testing and project planning will enable you to use your soil improvement budget as efficiently as possible.

Setting Objectives

This is where any good plan starts. What are your goals for your garden?

If you haven't given this much thought, consider the reasons you took up gardening in the first place (there were probably several):

- Fun.
- Beauty.
- Exercise.
- Self-sufficiency.
- Healthier food.
- Outdoor recreation.
- Additional income.

- Reduced grocery bills.
- Better produce flavor.
- A connection with nature.
- An educational activity for the children.

Write down one to three broad long-term objectives that you have for your garden as a whole.

Next you will want to list several shorter-term goals (probably about three to five) that will help you reach these long-term objectives. These should be projects that you can complete in one growing season. While we are going to focus strictly on soil improvement in this book, your seasonal goals do not have to be soil-related. This is an excellent opportunity to think about how you want to improve the garden in general.

Seasonal goals you could set, depending on your gardening objectives, include:

- Build a new raised bed.
- Solve a drainage problem.
- Raise your own seeds for next year.
- Implement an intensive gardening method.
- Add annual flowers to your vegetable garden.
- Reduce your grocery bills by a specific dollar amount.
- Grow vegetables so delicious that your children ask for seconds.
- Increase the brix content of your produce to a specified amount.
- Successfully harvest a vegetable that has been giving you trouble.
- Sell a certain dollar amount of produce at the local farmers market.
- Extend your growing season so that you harvest at least weekly through the winter.

Be sure to write both your long-term and short-term goals down so that you can refer to them later. If you are keeping a garden journal, consider noting your goals toward the beginning.

Identifying Limiting Factors

Once you know what you're aiming at, it's time to identify precisely what has the potential to hold you back. This will help you use your soil improvement budget as efficiently as possible.

Obviously, we don't want to end up with plants that are forever limping along, always merely surviving and never really thriving. We want plants that are vibrantly healthy, because these are the plants that will nourish us most abundantly.

But to avoid discouragement or overspending, it is necessary to keep in mind that soil improvement is a process. For badly abused soils, it is a long process (which is why some opt for alternatives, as we shall see in a moment). You will not achieve optimal results in the first year, and perhaps not in the first two or three years. But by really homing in on the major limiting factors, you will improve the soil as rapidly as possible.

Soil-related limiting factors can include:

- Shallow soils.
- Contamination by toxins, diseases, or pests.
- Poor balance of air and moisture.
- Energy imbalances (pH being the most common).
- Nutrient imbalances.
- Absence of organic matter and humus.
- Weak soil community.

This list is roughly organized in the order in which you will have to work to improve your garden soil. However, the factors toward the bottom of the list are by no means less important. Quite the contrary! A healthy soil community is absolutely crucial for producing tasty, nourishing produce. But you cannot have a soil community without the proper organic matter and nutrient levels. These nutrients, in turn, will not become available to plants without the proper energy balance. The energy balance is irrelevant if your soil is either too dry (too much air) or too wet (too much moisture). And no matter how much work you put into drainage and aeration, it is all in vain if you have heavy metals polluting the soil in toxic concentrations or if the soil is too shallow to support plant roots.

Therefore, a good place to start when looking at limiting factors is at the top of the list. Go back to your notes on the soil tests you performed and notice where the first breakdown occurs.

Table 6 Identifying Limiting Factors

Problem	Examples
Shallow soil	Erosion, no topsoil, hardpan, rocky soil, bedrock less than a foot below the ground
Contamination by toxins, diseases, or pests	Heavy metals, pesticide residue, black walnut, plant diseases, insect pests, moles, aggressive weed growth
Poor balance of air and moisture	Waterlogging, gray color, anaerobic smell, compaction, percolation too fast or too slow
Energy imbalance	pH too high or too low, sour smell, Albrecht test results out of balance
Nutrient imbalance	Symptoms of a deficiency or excess of any nutrient, deficiency or excess on NPK or lab test
Absence of organic matter and humus	Soil too clayey or sandy, poor structure, pale color
Weak soil community	No earthworms, organic matter not decomposing

Also look for repeated themes that crop up over and over again in your test notes. Glaringly obvious problems will need to be addressed sooner rather than later.

Note that there may be other limiting factors in your garden, such as excessive shade, that you will want to take into consideration, but in this book we will stay focused on the soil.

Once you know what the limiting factors are in your garden, jump ahead to the appropriate chapter and start fixing things:

- Chapter 3 covers building up shallow soil.
- Chapter 4 addresses toxins, diseases, and pests.
- Chapter 5 tells you how to balance the air and moisture in your soil.
- Chapter 6 provides insight on soil energy balance.
- Chapter 7 discusses the many nutrients plants need to stay healthy.
- Chapter 8 is about building organic matter and humus.
- Chapter 9 offers advice on caring for the soil community.

And if your soil is already in reasonably good health, Chapter 10 will tell you how to keep it that way.

But wait! What if you have exceptionally poor soil? Is it still worth improving?

Alternatives to Improving Problem Soils

Some soils, such as polluted urban soils, are so severely damaged that improving them can be dauntingly expensive and time-consuming. Other soils are simply unsuited to the purpose at hand; for example, blueberries have specific pH needs that are hard to meet in many parts of the country.

Mel Bartholomew of the best-selling book *All New Square Foot Gardening* noted that it takes about seven years to bring soil to peak productivity. He also noted that most people move within seven years, making soil improvement a waste of effort in many cases.

That said, soil improvement has two enormous benefits that may repay the work. First, it is the key to growing nutrient-rich produce that nourishes us and helps us achieve maximum health. Second, it is an important part of land stewardship.

Ultimately, only you can decide if improving your soil is worth the effort. If you are concerned about toxins in the soil that your family might consume, you may be better served to consider an alternative. Likewise, if you need to grow root crops in shallow soil, you may want to look at your other options.

Raised Bed Gardening

One way around the poor-soil conundrum is to garden in raised beds. The idea is that you will build a bed out of wood, stone, or another material to be filled with better soil from another source. The soil could be bagged topsoil from the garden center, or it could come from another place on your property.

In fact, Mel Bartholomew's square foot gardening method could be considered a variation on raised bed gardening. After building square boxes to contain the soil, he filled them with his own blend of equal parts vermiculite, peat moss, and finished compost.

As with most things, raised beds offer both advantages and disadvantages.

Advantages of raised beds include:

- **Attractive appearance.** A well-built raised bed looks much tidier than most row gardens. Select materials that appeal to your aesthetic sensibilities for a customized landscape.
- **Opportunity to start with good soil.** Do you have severe problems with your soil, such as heavy clay, skewed pH levels, low levels of NPK, or black walnut toxin? Filling a raised bed lets you purchase or mix perfect soil.
- **Fewer weeds and rodents (depending on construction).** Tired of fighting the weeds? Build a raised bed and line the bottom with cardboard or landscape fabric before adding soil. Are voles or gophers a problem in your area? Put a screen in the bottom of the bed to fence them out.
- **Early seed starting.** The elevated soil in the raised bed will warm up quickly in the spring, much quicker than the soil in a conventional row garden. You should have a couple weeks' head start on the growing season.
- **No trampling.** Perfect for those with high-traffic gardens. Raised beds protect soil from compaction by pets, small children, and others.
- **Good soil drainage.** Raised beds drain well, preventing waterlogging in wet climates.
- **Minimal erosion.** While some soil could get out through cracks in your raised bed, erosion will be pretty minimal thanks to the retaining walls. If you have no choice but to garden on a steep slope, raised beds are a great choice.
- **Ergonomics.** Raised beds are easier on the back, important to consider for the elderly or achy gardener.

But there are disadvantages:

- **Extra cost of materials.** You can't just till and go—you will actually have to build something, and that requires parts. (Be careful not to build with materials that will leach toxins, such as some plastics and treated lumber.)

- **Extra initial work.** Not interested in working with wood or stone? Maybe raised beds aren't for you.
- **Need to bring in soil.** You will have to either buy soil or dig it up from some other part of your property to fill the raised bed. This will cost either money or effort.
- **Tillage difficulties.** It's hard to get a rototiller into a raised bed. In fact, depending on the raised bed, it may be downright impossible. Raised bed gardening presents a good opportunity to go no-dig, which has many advantages as we will see in later chapters, but if you are sure you need to till you may want to garden using a more traditional method.
- **Disappearing soil.** Unless you are actively building soil, the overall soil level will drop over time as settling occurs and organic materials break down. This can happen in a ground-level garden, as well, but the effect is particularly startling in a raised bed. Attention to soil care will still be necessary.
- **Summer soil baking.** An elevated bed is more exposed to sun and evaporation. This means that you will be watering your garden more, and you may need to start cold-season plants a little earlier to compensate for the faster rise in soil temperature. Shade cloth will also prove to be a good investment in this scenario.

As you can see, raised bed gardening does not make you completely exempt from the task of soil improvement. It gives you a leg up by starting you off with soil of reasonable quality.

Container Gardening

Container gardening is another excellent alternative ideally suited to small spaces.

Container gardening advantages include:

- **Apartment-friendliness.** Five-acre homestead? Postage-stamp yard? No land at all? No problem! You can still garden—in containers!
- **Suitability for physically challenged gardeners.** If you love plants but find crawling around on the ground painful or difficult, you can create a container garden on a table or other elevated surface.
- **Mobility.** Plants can be brought indoors, moved to shadier areas, or switched around for a whole new landscaping look.
- **Convenience.** Even if there is nothing preventing you from gardening conventionally, you still might love having a few tomatoes and herbs right by the kitchen door. As an extra bonus, handy access to your containers will make your gardening chores quick and easy.
- **Fewer weeds.** As long as you fill your containers with weed-seed-free soil, the only weeds you will have to worry about are those that blow in, like dandelions. But even these will be easy to pull in the nice, loose soil you will have in your container.
- **No rabbits.** Are rabbits a continual problem in your area? That's easy to solve by elevating your container garden.
- **Unique look.** Gardening with containers can showcase your creativity. Whether your style is traditional or unconventional and upcycled, the combinations of containers and plants you put together will make your landscape look like none other.

There are a few disadvantages to container gardening, however:

- **Cost.** First you will have to buy containers—probably quite a few. Then you will have to buy soil. (Using soil dug up from the backyard is not recommended because it will likely not drain adequately when placed in containers, and it will probably also contain weed seeds.)

- **Fewer selections.** Not all plants will thrive in containers. Pumpkins are too large and sprawling to do well in containers, for instance. Root crops, such as carrots, can be grown in containers, but the container must be at least 12 inches deep.
- **Less vibrant soil.** Both nutrients and microscopic life forms will be in short supply in your new soil. You will need a quality organic fertilizer, and you will need to start over with new soil periodically.
- **Frequent watering.** Potting soil drains quickly, particularly when in small, elevated quantities in direct sunlight. Watering your container garden may be a daily (or twice-daily) task.

There are special considerations associated with container gardening, but none are deal-breakers if you are prepared for them. If for any reason conventional gardening is not a good fit for you, then container gardening might be worth a try.

Other Options

Raised-bed gardening and container gardening are the most common alternatives to having a designated garden plot at ground level. However, there are other options:

- **Aquaponics.** Similar to hydroponics, but with water from a nearby fish tank pumped in to fertilize the plants. Allows plants to feed somewhat more naturally than they can with hydroponics. However, it is expensive and high-maintenance because of the additional equipment and care the fish require.
- **Core gardening.** A variation on raised-bed gardening in which a trench is dug along the middle of the raised bed, filled with straw, and watered aggressively to start the decomposition process. Plants are grown in the trench. Very rich soil is achieved this way, but it does take some care in the beginning to get the core off to a good start.

- **Hugelkultur.** German for "mound gardening." Involves building planting mounds with logs, tree branches, and twigs to be filled with organic matter and buried in soil. Results in an incredibly rich medium for plant growth, but is very complicated to build.
- **Hydroponics.** Growing plants without soil. Weed-free. Requires fertilizer, pumps, timers, and artificial lighting, making it extremely expensive.
- **Keyhole gardening.** Similar to raised-bed gardening, but the bed is circular with a compost pile in the center and an aisle into the bed for easy access. The central compost pile is a superb way to maximize soil nutrition, but the setup is precise and it does require diligence to maintain.
- **Pallet gardening.** The horticultural response to the pallet project craze. Landscape fabric is used to convert a pallet into a container. Pallet gardens can be either horizontal or vertical for more efficient use of space. However, this method is not suitable for deep-rooted plants.
- **Soil bag gardening.** It doesn't get much easier than this—buy a bag of potting soil, cut an opening into the package, and plant directly in the bag. There's not much that can go awry here. Do keep in mind, however, that your produce is only as good as the soil you purchase, and that the plastic will eventually start to degrade.
- **Square foot gardening.** As already discussed, square foot gardening is really a variation on raised-bed gardening. However, it is unique because of the hallmark square-foot grid used to maximize efficiency. Relies on a purchased blend of vermiculite and peat moss, as well as compost.
- **Straw bale gardening.** A straw bale is "conditioned" with water and fertilizer to start the decomposition process, and then used as a planter. Inexpensive and space-efficient, but requires very frequent watering and plenty of fertilizer to meet plant nutrient needs.

- **Tire gardening.** Exactly what it sounds like. Inexpensive and a great choice for root crops. However, you may or may not like the way it looks.

As you can see, all methods have their unique pros and cons. The key is to choose a method that will best meet your needs, while taking care to make sure that plants receive the nutrients they need to produce health-giving vegetables.

Chapter 3
Building Soil Depth

If you have shallow, rocky soil, the first thing you will need to do is build it up, because without sufficient soil depth your plants will not thrive, no matter how favorable the nutrient profile is. Obviously, root crops, especially carrots, will become deformed trying to adapt to shallow soil. But many other plants will struggle, as well. For one thing, shallow soils dry out quickly, making plants prone to wilting in the summer heat. For another thing, shallow soils do not permit plants to anchor themselves firmly in windy areas.

Some smaller plants, such as lettuce and peas, can survive in soils that are six inches deep. Root crops and larger plants, like sweet corn, will benefit from soil that is at least one foot deep.

In this chapter, we will learn how to build up shallow soils, remove rocks, and break up hardpan layers that are preventing plant roots from digging down deep. We will also look at options for dealing with extreme situations where there is no topsoil to start with, such as on old excavation sites.

Preventing Erosion

The first thing to do is to prevent soil from disappearing after a rain or wind event. The key to preventing erosion is to make sure that the soil is never exposed. There must always be something covering it, whether it is vegetables, mulch, or a cover crop. Even weeds are better than bare soil.

Therefore, preventing erosion in the garden is actually quite easy. All you have to do is get something on top of the soil.

Building Up Shallow (or Nonexistent) Soils

Shallow soils make it harder for plant roots to probe deep enough to find adequate nutrition and moisture. Most plants need at least six inches of soil, and larger garden plants, such as corn and tomatoes, as well as root crops, benefit from a good foot of soil depth. And deeper never hurts.

In a few cases, you may be facing a site without any topsoil. This is an unfortunate situation indeed, and should not occur unless your garden site was previously subjected to excavation, construction, or some other severely disruptive activity. Ideally, you will choose a more favorable location for your garden. If this is not possible, you may seriously want to consider an alternative to soil improvement, as already discussed. If you are determined to improve your soil in this situation, however, it can be done, just perhaps with a little more expense than is ideal.

If soil depth is a limiting factor in your garden, you will likely have to bring in soil from an outside source to build up the garden site. If you happen to live on a larger property, you may be able to supply your own soil (need a pond?). Many backyard gardeners will have to make a purchase, however.

Be very careful when purchasing soil from an outside source. Much of what is sold as "topsoil" is extremely poor-quality. Never buy topsoil sight unseen. When you inspect soil prior to purchase, evaluate it as you did your garden soil, checking for color, smell, and texture. And while you are at it, consider purchasing peat moss and a few bags of several different types of compost to start your garden off with organic matter and a good mix of nutrients.

Over time, if you continue to work through this book, you will be building up your own soil by adding mulch, compost, and the like. For soils where the depth is not an immediate barrier to plant growth, this slower approach is ideal because there is less risk of introducing contaminants.

Breaking Up Hardpan

Hardpan can be the bane of a gardener's existence.

Once a layer of hardpan forms, no water can penetrate the soil beyond. Any moisture that reaches the hardpan layer will simply run off.

Furthermore, roots cannot enter the hardpan, either. Once a probing root tip feels the hardpan, it will turn aside, directing root growth horizontally. Or, worse yet, the plant will give up growing roots altogether.

This phenomenon can have two detrimental effects:

- Shallow-rooted plants with poor drought tolerance.
- Wet beds that collect water on or near the surface, drowning plants.

Hardpan can form in different ways. Sometimes it is a natural occurrence, while other times it is a manmade event.

Deposits are a major cause of hardpan in some areas. Hardpan-forming deposits can include:

- Sodium.
- Dissolved silica.
- Rust/limestone combinations.

Clay soils are particularly prone to hardpan problems because their particles carry strong negative charges, thus inviting bonding with other substances. Acidic soils, which have a positive charge, suffer from similar problems.

Any factor that tends to promote soil compaction can also create hardpan. In the garden, two common factors include:

- Heavy foot traffic (including from pets).
- Repeated tilling, particularly at the same depth every year.

However, soils can form hardpan without help from compaction.

If you have discovered that hardpan is a problem in your soils, never fear. The solution will take time, but there is hope nevertheless.

Your goal is to restore the capillary structure to the soil. This can be done using several techniques:

- Physically break up the hardpan layer. In the garden, this is typically done with a broadfork, which is a U-shaped tool with tines on the horizontal bar to loosen and aerate soil without inverting it.
- Plant a strong-rooted cover crop. The ideal crop for this job will have the strength to force its roots into the hardpan and start to break it up. Good choices for deep compaction include chicory, tillage radishes, sweet clover, medium red clover, and fall mustard crops. For hardpan closer to the surface, consider annual ryegrass, hairy vetch, white clover, and forage turnips.
- Work to build the level of organic matter in the soil. Organic matter separates soil particles, preventing compaction. We'll delve into this topic more in Chapter 8, but for now consider forking in compost, peat moss, shredded leaves, or dry straw.

Again, curing hardpan takes time. The more aggressively you pursue your soil improvement program, the faster you will begin to see results.

Dealing with Rocky Soils

How to deal with rocky soils depends on how bad the situation is. As you build soil and add organic matter, many of the rocks will be buried, and the moisture- and nutrient-retaining properties of your soil will improve. In the meantime, pick out rocks as you go through your normal routine of planting, weeding, and harvesting.

For small spaces and raised beds, you may have the opportunity to sift your soil. Soil sifters can be purchased, but they are easy to make, as well. Build a wooden frame to fit over a garden cart or wheelbarrow, and then staple galvanized hardware cloth to the top of the frame. Use it by hoeing soil over the hardware cloth; the soil will fall through the mesh into the cart, while the rocks and large dirt clods will be scraped off onto the ground. Alternatively, you could build a smaller frame with handles and staple the hardware cloth to the bottom, allowing you to fill the box with soil and shake the sifter by hand over a cart or garden bed to release the good soil.

A final suggestion is to grow your root crops and deep-rooted plants in deep containers while you are working on correcting the soil situation in the rest of your garden. You should be able to keep growing shallow-rooted crops, however, such as lettuce. Just be aware that these plants will need frequent watering, as rocky soil retains moisture poorly.

Chapter 4
Eliminating Toxins, Diseases & Pests

Is there evidence of residual toxins in your garden? City lots are sometimes contaminated by heavy metals, and bringing in grass clippings from an herbicide-treated lawn can introduce enough chemicals to kill your vegetables.

What about the pests and diseases common to many gardens? Unfortunately, you may not know what lurks beneath the soil in the form of fungi, bacteria, or larvae until you start gardening.

But once you do identify a problem, it's time to take action. This chapter will walk you through the clean-up process.

Toxins

There are several toxins that can affect garden soil. Heavy-metal pollutants, herbicides, and toxic plants are all potential problems.

Heavy Metals

Heavy metals can found in a variety of locations. They may drift in on the air in polluted cities, or they may lurk in soils where old-fashioned pesticides were previously used. They might leach out from old buildings, too. While all plants do not necessarily absorb all types of heavy metals, you can be exposed to heavy metals in your garden any time you come into contact with the soil.

Take the following steps to limit your exposure to heavy metals in the garden:

- Site your garden as far away as possible from potential sources of contamination, such as old buildings.
- Keep the soil covered at all times (e.g., with mulch).
- Keep tillage to an absolute minimum, preferably going no-dig if possible.
- Wear gloves, long pants, and closed-toe shoes when working in the garden.
- Wash your hands after working in the garden.
- Wash all produce thoroughly under cold running water before eating, cutting, or cooking.

Of course, if you are living with a known heavy metal issue in your soil, the best thing you can probably do is switch to container gardening for safety.

Herbicides

Time is a major factor in dealing with herbicide residues. Every herbicide has an estimated half-life. Some vaporize in short order, while others persist for months on end. Herbicides licensed for homeowner use break down quickly; those used for agricultural purposes may linger longer. Therefore, it helps if you know what herbicide you are dealing with.

Another factor is the source of the herbicide. Some herbicides lurk as residues in everything from lawn clippings to horse manure to commercial compost. Others are harder to avoid because they drift.

A thick treeline can help with herbicide drift, as long as you are willing to sacrifice the health of the trees for the cause.

Finally, if pesticide drift is a recurring problem in your area, you may want to see if you can come to an amicable resolution with the perpetrator. Please tread lightly, as pesticide drift has become a hot-button issue in rural America. Be polite and friendly in your speech, avoiding harsh accusations. The local farmer probably feels that he needs to use the herbicide to stay in business, but he might be willing to give you fair warning before spraying so that you can cover your plants with plastic. If the spraying is conducted by the local or county government, they may have an alert system already in place.

In a worst-case scenario, you should have some recourse by contacting your state's department of agriculture.

As for dealing with herbicide residues already contaminating your soil, patience is a virtue. Most residues will break down within a month, a couple years at the worst. While you wait, consider these options:

- Grow only plants that are not susceptible to the herbicide.
- Grow sensitive plants in containers.

Meanwhile, to help the healing process, consider tilling the affected area and planting a cover crop. Watering the soil thoroughly while the cover crop is growing will also help.

Toxic Plants

Some plants exude toxins in a bid for ecosystem supremacy.

Common plants that may attempt to hinder your gardening efforts include:

- Gladiolus.
- Sunflower.
- Black walnut.

Avoiding toxins from most plants, such as gladiolus and sunflower, is relatively easy—just don't grow these flowers near vegetables.

Black walnut trees are a little harder. They exude a toxin called *juglone*. Gardens planted near black walnut trees or rootstocks frequently fail to thrive due to this toxin. Many plants can be affected, but tomatoes and potatoes show the greatest susceptibility. Juglone enters the soil from the roots, bark, leaves, and nut hulls of the walnut tree. It cannot move far, as it is not water-soluble, but it impacts nearby garden plants by interfering with their respiration. This weakens the plants and eventually causes their death. There is no cure for walnut toxicity. If it becomes a problem, the garden must be relocated or moved into raised beds. If you can, choose a garden site away from walnut trees. On small properties, this may not be an option, but you still have two other possibilities to consider:

- Choose only plants tolerant of juglone (e.g., beans, corn, squash).
- Grow all plants in raised beds, regularly swept clean of leaves and nuts.

Even if your garden is in an ideal location, never use black walnut leaves, bark, or chips as mulch.

Diseases

Many garden diseases are soil-borne. Once a plant becomes infected, the pathogen takes up residence in the soil or in plant debris in the soil, lying dormant for years if necessary to find a new host.

Sound garden hygiene practices will prevent most soil-borne diseases:

- Buy seeds from reputable sources.
- Grow your own seedlings rather than purchasing them.
- Fill seedling containers with quality seed-starting mix, not garden soil.

- Disinfect seedling trays and containers after use.
- Store seedling trays and containers in a dry area to discourage fungal growth.
- Plant in a location with plenty of sunlight.
- Space plants far enough apart for good airflow.
- Take steps to improve soil drainage, if necessary.
- Rotate garden crops.
- Wash your hands before touching your plants.
- Regularly clean garden tools, plant stakes, and other items that come into contact with soil.
- Avoid feeding plants excess nitrogen, as it promotes rapid growth of soft, injury-prone tissue that gives pathogens easy access.
- Do not work with plants when wet.
- Do not use sprinklers in the garden, as this wets the leaves of the plants and fosters fungal growth.
- Keep weeds and insect pests in check.
- Support plants to keep their fruits and leaves off the ground.
- Prune climbing plants regularly to remove damaged tissue and keep growth under control.
- Ferment seeds from tomatoes, cucumbers, and melons before storing them.
- Remove plant debris in the fall.

To treat plant diseases, you will want a gardening encyclopedia. Keep in mind, however, that it is rarely worthwhile to treat annual plants for diseases; the better bet is to remove and destroy the diseased plants to keep them from contaminating the soil and thus infecting the next season's plants.

If you have a severe and persistent disease problem in your garden, you may not be able to continue growing susceptible plants and plant varieties. In that case, consider container gardening as a way to keep raising your favorite plants.

Pests

Some of the most annoying pests in the garden can be those that live in the soil. Cutworms and grubs wreak havoc as they eat, while mice and moles leave destruction in their wake as they travel from one place to another. And then there are the weed seeds!

So what do you do if your garden is overrun with these unwelcome visitors?

Insects and Their Kin

A big part of keeping insect pests in check is keeping plants healthy. In nature, weak organisms are typically attacked first. In the animal kingdom, predators see and smell the weakness of their prey. Gardeners have long suspected that diseased or dying plants put out some sort of "distress signal" that attracts scavenging insects to the scene.

So what can you do to specifically deter garden pests? Here are some tips:

- **Water your plants.** Insect pests typically prefer to attack stressed plants, so keep your garden healthy. As temperatures rise, make sure all your fruits and vegetables are adequately watered.
- **Remove dead and dying plant material.** Bugs like stressed plants, remember? Most garden plants can tolerate a little pruning when needed. Any time you see dead plants, diseased foliage, or rotting produce, dispose of it at once.
- **Introduce predators.** This could be a couple of well-behaved guineas, or it could even be more bugs. Although you can buy a swarm of ladybugs, sometimes the best bet is just to relocate a predator you already have on your property. For instance, the garden spider in your garage might actually enjoy living outdoors better once he gets used to the change. Most helpful predators, however, will naturally feel welcome in a diverse garden free

from toxic chemicals. Also consider setting up birdbaths and houses to attract birds to your garden.

- **Mulch selectively.** Mulch is very useful for keeping the soil moist and the weeds under control, but sometimes it provides hiding places for your unwelcome guests. Squash bugs, for instance, love to lurk in the privacy of a thick straw mulch. If you know you have a pest with an affinity for mulch, spread the straw and leaves a little thinner and water more frequently.
- **Mow the surrounding area.** Keep the grass short around your garden. Don't leave inviting hiding places to serve as bases for insect attacks.

Notice that keeping plants healthy and destroying the haunts of harmful bugs are crucial to preventing a major onslaught.

You may still see a few bad bugs and minor plant damage, but you should manage to avoid an infestation with the help of these tips. Any pests that do appear on the scene can be controlled naturally by hand-picking or with a generous dusting of feed-grade diatomaceous earth.

Subterranean Mammals

An effective way to exclude moles and other underground pests is to build a raised bed 12 to 18 inches deep and line the bottom with galvanized hardware cloth before filling the bed with soil. Hardware cloth is tough enough to stop most rodents and moles, while the galvanized coating protects the metal from rusting for years, often decades.

However, it is also worth considering why these pests are swarming into your garden in large numbers in the first place. Rats, mice, voles, and the like may be attracted by brush piles, garbage, and other inviting sources of food and shelter, which should be removed. Moles are more likely attracted to large quantities of grubs, which can multiply if you use too much water and fertilizer in your garden or on your lawn.

Other ideas for controlling burrowing vermin naturally include:

- Make compost in fully enclosed bins.
- Keep the grass mowed in and around the garden area.
- Interplant strong-smelling herbs such as basil and garlic liberally throughout the garden.

Weed Seeds

Weed seeds take a great deal of time to combat.

A simple way to prevent weeds from becoming a problem is to mulch heavily and never till the soil. This will keep most of the existing seeds buried, and those near the top will largely be prevented from germinating by the mulch. Cardboard or newspaper under straw is even more effective as a weed barrier.

Can tillage be used to control weeds? Certainly! Keep in mind, however, that every time you till you are bringing up a fresh crop of weed seeds from underneath the surface. The success of tilling to control weeds is heavily dependent on timing. You won't be able to grow any vegetables in the plot while this process is going on. Allow a crop of weeds to germinate, and till it under before it gets established. Then let another crop germinate, and till that. Continue until you are satisfied with your results or you simply decide to use the gardening space.

Another option is to grow a thick cover crop to smother the weeds. Cover cropping will be discussed in more detail in Chapter 8. For now, suffice it to say that a good, thick stand of winter rye is superb for fighting weeds.

In extreme cases, solarization is an option. This involves covering the soil with clear plastic so that the sun can cook it. This will kill off any seeds close to the surface, but it will also devastate the soil community.

Of course, you must also avoid introducing new weeds to your garden in the future. This can be at least partially achieved by paying attention to the following:

- Buy seeds from quality sources.
- Use seed-free mulches (straw instead of hay, for instance).
- Pay attention when mowing or weed-eating around the garden to avoid flinging seedheads into the plots.

If you do end up with weeds, it's important to remove them before they produce seed. Pulling them by hand is the most effective way, but cultivation may be necessary in a larger garden.

Chapter 5
Balancing Soil Air and Moisture

If you will recall from earlier in this book, healthy soil is about 25% air and 25% moisture. If there is too much air, the plants will dry out. If there is too much moisture, the plants will drown.

Fortunately, whichever problem you need to deal with, correcting it does not have to be complicated.

How to Aerate Soil

Soils tend to need more air either when there is too much moisture or when there is not enough space between soil particles (poor soil structure). The lack of space between particles typically occurs when fine-textured soils, particularly clay, become compacted. Waterlogging occurs when excess moisture cannot drain quickly enough, perhaps because of a layer of hardpan beneath the topsoil.

Improving a soil aeration problem can be accomplished in four steps:

1. Improve soil drainage.
2. Break up surface compaction (dealing with subsurface hardpan was addressed in Chapter 3).
3. Improve soil structure.
4. Protect the soil from further damage.

Improve Soil Drainage

Improving soil drainage can be quite a project, but when it comes to dealing with heavy, fine-textured soils, it is vital to keeping your plants healthy.

The first step to ensuring good soil drainage is choosing a good garden site. You really don't want your garden looking like a pond in spring, so it's best to avoid low spots. A gentle south-facing slope can help prevent waterlogging problems.

If you have no choice but to plant your garden or orchard in a poorly drained location, you will need to build the soil, certainly with the addition of generous amounts of organic matter and possibly with more topsoil. Of course, in truly severe cases, your best bet may be to avoid the problem altogether by planting in containers or raised beds.

Once you have improved your garden site to the point where it generally doesn't resemble a bog, the rest is fine-tuning. Keep adding organic matter until your soil reaches the ideal texture and passes the percolation test described earlier.

Break Up Surface Compaction

The first step to aerating soil is to break up any compaction that may be near the surface. There are two major ways to go about this.

The first method is to plant a cover crop. Cover crops break up compaction by penetrating soil with sturdy root systems. They also help prevent compaction from reoccurring by adding organic matter to the soil, which puts space between finer soil particles. The primary drawback of using cover crops to relieve soil compaction is that at least a portion of your gardening space will be unavailable as the cover crop grows. Also, after you mow down the cover crop, you may need to let it decompose a bit before planting finer seeds in it, as coarse debris may prevent the seeds from making good soil contact and thus hinder germination.

Some cover crops are better suited to breaking up compaction than others. The best cover crops for this purpose are those with deep roots. Choosing and growing a cover crop will be covered in more detail in Chapter 8.

For quicker results, you can use a tool called a broadfork. This handy tool is U-shaped, with two vertical handles that you hold on to and a horizontal bar on which you put your weight. This horizontal bar is equipped with tines that poke into the soil and loosen it up while doing minimal damage to the soil structure or the soil community.

Improve Soil Structure

Soil structure is improved with the addition of organic matter to separate soil particles. This will be covered in more detail in Chapter 8.

Protect the Soil

The key to keeping soil from compacting in the first place is to avoid trampling it. This means no vehicles in the garden plot, obviously, but it also implies minimizing foot traffic.

Ideally, you will mark out clear, permanent paths throughout the garden. These paths should not move around from season to season—foot traffic should be restricted to designated places to prevent compaction of the growing area. Clearly marking the paths will also discourage people from stepping off the paths. You will need to teach your children to keep to the paths, and pets should be kept out of the garden altogether.

These paths don't have to be elaborate, just wide enough that you won't trip yourself up. Two feet wide is about right—narrower paths are hard to navigate, while wider paths increase the area that must be weeded without increasing the harvest. That said, if you are seriously klutzy or have difficulty maintaining your balance, you may benefit from paths that are three feet wide. Make sure you give yourself plenty of walking space in these key areas:

- Between rows.
- Between raised beds.
- Between boxes or blocks in a square foot garden.
- Between the fence and the plants or beds.

How you build your paths will depend on the circumstances. Paths through the row garden can be mulched with straw to prevent them from turning into a muddy mess when it rains. If you want something longer-lasting, consider wood chips or rock stepping stones.

To further reduce the risk of stepping into a bed, make sure that you don't make it so wide that you can't reach the middle. Ideally, your beds should be separated from fences by a path, but if you must put a bed flush against the fence, make it narrower to compensate. Two or three feet wide is probably ideal in this situation, while a free-standing bed should be about four feet wide.

Another key to preventing soil compaction is to avoid working clay soils when they are wet, as this is a good way to destroy soil structure and create bricks.

Here is an easy way to tell if your soil is safe to work:

1. Evaluate the soil surface conditions. If there is still snow and ice on the ground, don't bother. If there is muck, ooze, and standing water on the ground, especially don't bother.
2. Use a trowel to dig up a soil sample about the size of your palm. If the ground is too frozen to get your trowel into, it's not warm enough yet. If any considerable quantity of mud sticks to your trowel, the soil is too wet.
3. Gently mold your soil sample into a ball and give it a good poke. If water oozes out or it sticks together like clay and requires some force to break up, it is too wet. If it is moist but crumbles apart

readily when poked, it is just right for planting. If you can't make a ball with the soil, it's dry; you can still plant, but you will need to water generously.

A bonus tip for the first spring planting: Pulling back the mulch to expose the soil just for a day or two can warm it up and dry it out quickly. As long as you replace the mulch within a day or two, you shouldn't do any real damage.

Finally, you may want to consider no-dig gardening as a way to aerate the soil. This one may come as a surprise to you. After all, tilling is supposed to be the way to loosen up the soil in the spring. However, at the farthest depth that the blades can reach, they actually stop turning the soil and start packing it down. If you till at the same depth every year, the soil immediately below the tillage zone turns into hardpan. No-dig gardens will not suffer from compaction as long as soil health is properly attended to and clearly defined footpaths are provided to avoid trampling the soil.

How to go no-dig? We'll discuss that more in Chapter 9.

How to Improve Soil Moisture Retention

If you live in a drier climate, or if you have sandy soil that tends to drain too quickly, you may need to help your soil retain moisture longer so that your plants can thrive.

The most important thing you can do in this regard is to continually strive to build the level of organic matter in your soil (more on this to come in Chapter 8). Organic matter acts like a sponge, filling the spaces between coarse soil particles and trapping water in their midst.

The second most important thing to do is mulch. Mulch protects the soil from rapidly drying in the sun and wind. Without mulch, you may have to water your entire garden every day in the summer. With mulch, you can water less frequently, promoting deeper root growth that will in turn make your plants more drought-hardy.

A final step you can take is to form shallow bowls in the soil around new transplants to trap water. This is not a tremendous soil moisture reservoir, but it makes a world of difference when dealing with those delicate seedlings.

Chapter 6
Charging Up the Soil Battery

On to a fascinating and all-too-often neglected topic—energy and how it flows through soil.

Soil energy is dismissed by some as superstitious humbug, but it is actually quite straightforward. We know that electricity does indeed flow through soil.

There is evidence that soil energy does indeed affect plant health. There is much that we don't know about this subject at present, but what we do know is nevertheless quite useful to maximizing plant vigor and therefore the nutritional quality of our harvest.

The Charges in the Soil

Did you know that soil has electric charges? Without these charges, soil would be unable to retain nutrients. So what do we need to know about the electric charges in the soil?

Why Does Soil Conductivity Matter?

Sand has low conductivity, while clay is highly conductive. The nutrients, salts, and moisture present in the soil also have an impact on soil conductivity.

It appears that there is an ideal balance when it comes to soil conductivity. Very high conductivity may suggest that there are high concentrations of salts near the surface of the soil, which hinders plant growth. On the other hand, very low conductivity may indicate that the soil is too sandy and therefore not retaining the moisture and nutrients plants need to resist the onslaught of pests.

The many variables that affect conductivity make it difficult to use a precise measurement as a means to troubleshoot soil problems, even for those who have the special equipment needed to take the measurements.

However, because conductivity correlates strongly to soil texture, with sand being poorly conductive and clay being very conductive and silt somewhere in the middle, you can automatically assume about what your soil conductivity will be given the soil texture. Likewise, if you tested your soil's cation exchange capacity (CEC), a high CEC will correlate with high conductivity.

For the time being, it is enough to remember that to grow plants need soil that is conductive, but not too conductive.

What is Cation Exchange Capacity (CEC)?

We have already touched on cation exchange capacity (CEC) when we discussed the Albrecht test. If you recall, CEC is the measure of the total negative surface charges in the soil. All atoms and molecules have a positive, negative, or neutral charge. Positively charged molecules are called cations, while negatively charged molecules are called anions. Because opposite charges attract each other, anions, such as clay and organic matter, attract cations, such as potassium and calcium. (Now you know *why* clay retains nutrients so well.)

Cations of interest to gardeners include:

- Nitrogen in the form of ammonium (not nitrate).
- Potassium.
- Calcium.
- Copper.
- Iron.
- Magnesium.
- Manganese.
- Zinc.

Therefore, CEC is important because it is a measure of how many nutrients your soil can hold on to. An ideal CEC is between 25 and 40. Soil with a CEC lower than the recommended range is probably too sandy with not enough organic matter, while a higher CEC indicates too much organic matter.

This is useful information to know, because there is a limit to the nutrients you can apply to a low-CEC soil at once, even if you are using natural materials. Applying too much nitrogen, perhaps in the form of chicken manure, is pointless and even detrimental on a low-CEC soil, because the nutrients will have nothing to bind to. Some of the nitrogen will bind to the available exchange sites, and the result will run off, possibly entering water sources.

On the other hand, it is possible to have too much of a good thing. If your CEC is too high, the soil will bind so tightly to its nutrients that plant roots cannot extract them again.

Adjusting CEC

Unless you happen to live in a peat bog, you will most likely deal with CEC measurements that are too low, rather than too high. These soils will benefit from the application of organic matter, as described in Chapter 8.

While excessively clayey soils will have a high CEC, a CEC that is too high for good plant growth (over 40) is more likely indicative of excess organic matter, as found in peat soils. If you are dealing with a peat-based soil, you may want to bring in some topsoil or finished compost to correct the problem. Also be sure to add more nutrients to the soil, such as manure. Even though your soil may be retaining plenty of nutrients, it is not giving them up readily, so a little excess will benefit your plants in this case. Likely the pH will need to be adjusted, as well (more on this in a moment).

Not sure what the precise CEC of your garden soil is? Pay attention to other indicators of too much or too little organic matter and adjust accordingly.

About pH

All gardeners have heard of the importance of pH. But did you know that pH is tied directly to the electrical charges we've been discussing?

What is pH?

The pH scale measures the concentration of hydrogen ions in a dissolved substance. Pure water can be used as a reference point; its pH is 7, or neutral. Any number below 7 is acid, and any number above 7 is alkali. An acidic substance has a greater concentration of hydrogen ions than pure water, while an alkali substance has a lower concentration.

It is interesting to note that pH also represents a relationship between the positive hydrogen (H) ion and the negative hydroxide (OH) ion. When the pH is neutral (7), hydrogen and hydroxide ions are in perfect balance. An acid has a preponderance of hydrogen ions, while an alkali has more hydroxide ions.

In other words, we could say that pH is a measure of the number of free hydrogen cations in the soil. The more hydrogen cations around, the fewer exchange sites will be available for other nutrients to bind to.

On a practical level, pH dictates what nutrients will be available to a plant. As the pH level moves along the scale, various minerals in the soil are locked and unlocked. Most of the nutrients that plants need to thrive are available when the soil is neutral to ever-so-slightly alkali.

Adjusting pH

pH is altered by adding amendments that tip the balance toward more acidic or more alkaline. Amendments that make the soil more acidic include:

- Peat moss.
- Pine needles.
- Sulfur or sulfate.

Amendments that make the soil more alkaline include:

- Lime.
- Wood ashes.

Adjusting pH is fairly straightforward. However, it is best not to attempt to make major pH changes without understanding the energy and nutrient balance of your soil as a whole. Adding peat moss or wood ashes from time to time likely won't hurt anything, but unless you have a serious problem you may want to avoid buying sulfur and lime. It is better to adjust the soil in a more holistic manner, by feeding it the organic matter it needs as described in Chapter 8.

Paramagnetism and How to Increase It

To state it in its most basic terms, paramagnetism refers to a weak attraction to a magnetic field. Soil is a paramagnetic substance.

Soils vary considerably in their alignment with the earth's magnetic field. Some have very little attraction to magnetic fields at all, but the most productive tend to be those with greater paramagnetism.

Interestingly, plants are diamagnetic, which means that they are slightly repelled by a magnetic field. Some gardeners believe that this property enables the plant to receive energy from outside sources, particularly paramagnetic soil. This attraction of opposites may create an efficient energy flow, increasing plant health and productivity.

The greater the paramagnetic field of a soil, the greater its productivity will be. A reading of 100 centimeter/grams/seconds (CGS) is the absolute minimum acceptable level. Excellent soil will test over 700, perhaps even over a thousand!

Paramagnetism is usually increased by the following good gardening practices:

- Balancing the ratio of calcium to magnesium (usually 10:1, although 7:1 is good on sandy soils).
- Increasing soil organic matter.
- Supporting the development of a healthy soil community.

For faster results, however, gardeners can amend their soil with paramagnetic rock dusts. These rock dusts usually come from granite or volcanic rocks, such as basalt. (In fact, the paramagnetic properties of volcanic rock could explain why volcanic soils are so conducive to plant vigor.)

Keep in mind that there is still much we don't know about paramagnetism and its influence on plants. Some sources claim that the paramagnetic properties of these rock powders stimulate root growth, while others state that remineralization is the only benefit derived. Either way, what we do know is that rock powders work.

Chapter 7
Balancing Soil Nutrients

Most gardeners are familiar with the importance of NPK, or nitrogen, phosphorus, and potassium. Those who pay attention to organic gardening literature may also be aware of the many micronutrients key to plant health, such as boron, manganese, and selenium. We have also addressed to some degree how the availability of these nutrients is largely controlled by pH and the flow of energy in the soil.

What is all too commonly overlooked is the balance between the various nutrients. A deficiency of one mineral, however seemingly insignificant, can render the plant unable to absorb other nutrients. Furthermore, an excess of one mineral can cause a deficiency of other minerals.

In this chapter, we will introduce the various plant nutrients, macro and micro, and offer insight into the balance plants need to stay healthy. Then we'll learn more about fertilizers, chemical and organic, as well as the pros and cons of the various choices.

The Macronutrients

NPK is an acronym you will frequently run across when dealing with plants. The three letters are merely the chemical symbols of three nutrients:

1. Nitrogen (N).
2. Phosphorus (P).
3. Potassium (K).

So why is there so much emphasis on N, P, and K? These nutrients are essential to plants because they serve vital functions.

Table 7 Macronutrients

Macronutrient	Functions	Deficiency Symptoms	Toxicity Symptoms
Nitrogen	Making chlorophyll; assisting plant growth and reproduction; synthesizing proteins	Slow, stunted growth; chlorosis; purple appearance on underside of leaves; purple stems; wilting and loss of older leaves; reduced yield	Wilting; excessively, bushy growth; burnt leaves; increased susceptibility to pests and diseases; susceptibility to drying out; absence of fruit
Phosphorus	Promoting fruit growth and ripening; enhancing root growth; providing vigor and stress resistance; enabling division and energy transfer	Vague symptoms except with severe deficiency; slow growth; unusually dark green color changing to red, especially on lower leaves; burnt leaf tips; leaf death; delayed fruit set; dead spots on fruit	Withering; generalized yellowing; iron deficiency; zinc deficiency

Potassium	Activating enzymes; regulating water pressure; building cellulose; enhancing produce quality	Symptoms appearing in older tissues first; wilting; stunted growth; increased susceptibility to disease and weather stress; scorched, curled leaf edges; chlorosis between leaf veins; purple spots on undersides of leaves	Nitrogen deficiency; calcium deficiency; magnesium deficiency; manganese deficiency

Nitrogen (N)

Soil nitrogen levels can easily be depleted by plants during periods of rapid growth. Some plants, such as corn, are particularly heavy feeders.

Also, soil microorganism health is key to making sure that any nitrogen in the soil is readily available for plant consumption. Therefore, in some cases, a nitrogen deficiency may be caused by insufficient soil life rather than an actual mineral shortfall.

Natural nitrogen sources include grass clippings, legume cover crops, and animal manure. Poultry manure is a particularly potent source, although bat guano and worm castings are useful, as well. Balanced compost will also contain nitrogen.

To prevent future deficiencies, consider taking the following steps to raise nitrogen levels and promote microorganism health:

- Improve soil drainage if necessary to avoid drowning beneficial microorganisms.
- Reduce or eliminate tillage to avoid soil disturbance.
- Rotate crops to reduce nitrogen depletion, preferably preceding hungry crops with legumes.
- Add cured compost or rotted manure to the soil regularly.
- Keep soil covered at all times to keep moisture levels stable.
- Use cover crop mixes containing legumes.

Nitrogen toxicity is typically the result of overly zealous fertilization. It's a very easy mistake to fall into because just a little nitrogen makes such a big impact on plant growth.

If you inadvertently apply too much nitrogen to the soil, an efficient way to lower the nitrogen levels without wasting it is to plant something that can absorb a great deal without adverse effects. Lettuce is particularly effective here, but corn or Brussels sprouts should work, too.

Barring that option, your next best bet is to add ample amounts of high-carbon organic matter to the soil. Microorganisms use up a great deal of nitrogen when breaking down carbon, so use this information to your advantage. This is a great time to try out a mulch made of wood chips, tree bark, sawdust, or shredded cardboard.

Finally, be sure to water generously. Not only will this help flush the excess nitrogen out of the soil, but your plants will appreciate the extra hydration at this time. Nitrogen causes salts to accumulate in the soil, impairing the plants' ability to stay hydrated.

Phosphorus (P)

Interestingly enough, most soils are deficient in phosphorus. This is because it is released slowly, yet plants have high demands for it. A healthy soil community is necessary to help plants take up phosphorus efficiently.

Phosphorus deficiency should be corrected with the application of a natural phosphorus source. One of the most potent natural sources of phosphorus is bone meal. Other sources include:

- Rock phosphate.
- Crushed granite.
- Composted leaves and grass clippings.
- Chicken or swine manure.
- Bat guano.

Boosting soil microorganism levels through good soil care practices will help prevent phosphorus deficiency in the future.

Phosphorus toxicity is typically caused by the overuse of high-phosphorus fertilizers (note that this includes manure). However, it can also occur if plants are watered with grey water from homes using phosphorus-based detergents.

Because excess phosphorus causes few problems in garden plants, reducing or discontinuing phosphorus usage will solve the problem.

Potassium (K)

Plants can easily become deficient in potassium simply because their requirements for this mineral are so high. Root crops in particular need a great deal of the K in NPK, but all crops use potassium at rates exceeded only by nitrogen.

Should you suspect potassium deficiency in your garden, apply greensand or hardwood ash for a quick boost. Sheep manure or compost containing fruit and vegetable waste are also effective options.

Plants generally do not experience outright potassium toxicity. However, when the soil potassium levels are excessively high, plants are prone to what is termed "luxury consumption." This in turn reduces the plant's ability to take up other nutrients.

This problem can be avoided and corrected by limiting the amount of potassium applied to the garden or field to what is required. Excess potassium can gradually be flushed from the soil by thorough watering.

The Micronutrients

Unfortunately, gardeners all too often focus exclusively on providing their plants with NPK and forget about the trace minerals (also known as *micronutrients*) that plants need. Plants rarely show obvious signs of a zinc, manganese, or molybdenum deficiency, although their hardiness and productivity is usually reduced. When humans eat food that has been grown on micronutrient-deficient soils for extended periods of time, however, their health may suffer. For this reason, it is important to take trace minerals into consideration.

Micronutrient excesses can also occur. These problems may throw the plant's nutrient intake out of balance, causing a host of other difficulties.

Table 8 Micronutrients

Micronutrient	Functions	Deficiency Symptoms	Toxicity Symptoms
Aluminum	Preventing plant disease; stimulating root growth	None	Failure to thrive; stunted growth; increased susceptibility to drought; pale color; calcium deficiency; soil pH below 5.0

Boron	Promoting ion exchange and nutrient transport; maintaining water balance; regulating plant hormones; promoting soil microorganism health	Pale or red new leaves with wrinkles; death of terminal growing points; susceptibility to larvae attacks; failure to set fruit; cracked, corky spots on fruits and tubers; lumpy strawberries; watercore	Stunted growth; pale leaf margins; leaf death between veins; reduced fruit yields
Calcium	Regulating pH; stimulating soil bacteria; regulating nutrient transport; serving as a structural component of cell walls; protecting plants from pathogen enzymes	Stunted, distorted growth; weak stems; death of terminal growing points; dark, curled leaves; leaf margin death; fungal disease; blossom end rot; bitter pit	Boron deficiency; magnesium deficiency; phosphorus deficiency; potassium deficiency
Chlorine	Splitting water during photosynthesis; regulating osmosis; regulating nutrient uptake; suppressing disease	Reduced growth; wilting; bronzed or mottled leaves; leaf curl and death; absence of normal cabbage smell	Reduced growth; bleached or scorched leaves; small, thickened leaves; premature leaf shedding; death

Cobalt	Serving as a constituent of several enzymes; promoting stem growth and leaf expansion; enabling legumes to fix nitrogen; providing drought resistance to seeds; forming vitamin B12	Vague symptoms; stunted growth; uniform paling, yellowing, or reddening of leaves; small root nodules; nitrogen deficiency in legumes	Stunted growth; pale leaves with discolored veins; leaf loss; failure to produce fruit; iron deficiency; death
Copper	Activating plant enzymes; manufacturing lignin; protecting plants from excess nitrogen	Cessation of growth; wilting; overall blue-green color fading to yellow; twisting and death of leaf tips; major reduction in produce yield and quality; death	Reduced growth and vigor; severe wilting; white areas between leaf veins
Iron	Forming chlorophyll; enabling respiration; assisting with nitrogen fixation	Stunted growth; leaf chlorosis, particularly between veins of younger leaves; centipede infestation; small flowers; failure to produce fruit	Wilting; bronzing or brown spots on leaves; reduced yield; manganese deficiency; death

Magnesium	Participating in enzyme reactions; serving as a central part of the chlorophyll molecule; regulating mineral absorption and metabolism; stimulating soil bacteria	Yellowing between veins and around margins of older leaves, perhaps with brown or purple spots; upward leaf cupping; susceptibility to larvae attacks; death	Yellowing or striping of leaves; reduced yield; calcium deficiency; potassium deficiency
Manganese	Activating enzyme systems; producing chlorophyll; metabolizing nitrogen and carbohydrates; producing lignin; promoting root crop health; synthesizing vitamins	Small, irregularly shaped leaves; checkered, striped, or uniform yellowing; death of discolored leaf parts; reduced yield	Symptoms mostly on older leaves; leaf cupping; darkened leaf veins; reddish-brown spots on leaves; burned leaf tips and margins
Molybedenum	Producing enzymes to enable nitrogen fixation, uptake, and metabolism	Downward leaf cupping; bleaching and death of leaf margins and areas between veins; reduced flowering; susceptibility to larvae attacks; reduced yield; nitrogen deficiency in legumes	None

Nickel	Promoting soil microbial health; enabling seed germination; improving iron absorption	Reduced productivity; chlorosis of new leaves; sparse foliage	Poor shoot growth; pale leaves; leaf death; deformed flowers
Selenium	Stimulating plant growth; providing stress resistance	None	Symptoms varying drastically by plant species; stunted growth; leaf chlorosis; withering and drying leaves; death
Sodium	Function unknown; may regulate water balance, enable carbohydrate metabolism, and enhance produce flavor	None	Wilting; stunted growth; susceptibility to drying out; scorched leaf tips and margins; reduced yield; unusually high soil conductivity
Sulfur	Transporting electrons during photosynthesis; enabling plant growth; enabling legumes to fix nitrogen; producing onion, garlic, and mustard aroma	Stunted, spindly plants; uniform chlorosis; susceptibility to larvae attacks; sharply reduced productivity	Nitrogen deficiency; low soil pH

Zinc	Participating in DNA transcription; activating enzymes that control plant growth; regulating water and carbon dioxide balance; enabling seed production	Shortened internodes; reduced shoot growth; stunted leaves with chlorosis between veins; dead spots on leaves	Symptoms subtle; stunted growth; dark leaves; reduced yield

Aluminum

Aluminum is common throughout the earth's crust, so a deficiency is highly unlikely. Instead, gardeners are more likely to deal with aluminum toxicity. Aluminum becomes highly available to plants as the soil pH drops below 5.0. Thus, aluminum toxicity in plants is usually caused by acidic soils. This may occur naturally or in gardens where the soil pH has been adjusted with aluminum, as is sometimes the case with acid-loving plants like blueberries and hydrangeas. Because of the risk of aluminum toxicity in surrounding plants and in humans who eat aluminum-fortified blueberries, adjusting the soil pH with aluminum is not recommended (try ammonium sulfate instead).

If you have determined that your plants are receiving excess aluminum, the first thing to do is to raise the pH, preferably to at least 5.5, so that your plants will stop absorbing so much. In most cases, this will be sufficient to correct the problem.

However, if you still have too much aluminum built up in the soil, perhaps from previous pH adjustments, consider growing tea or buckwheat as a cover crop for a few years. These species will help pull the excess aluminum out of the soil.

Boron

One of the most common micronutrient deficiencies worldwide is boron. Deficiency is most common in dicots on the verge of flowering, particularly strawberries, apples, grapes, pears, and walnuts. The problem is that plants must have access to a steady supply of boron at the root level, particularly before flowering. At the same time, boron availability is a tricky thing. When soils are wet, boron leaches out readily; when they are dry, boron is harder for the plant to absorb. Other factors that can contribute to a boron deficiency include:

- High air humidity levels.
- Low soil temperature.
- Sandy soil texture.
- High soil pH.
- Excess soil calcium.
- Severe nitrogen deficiency.
- Low organic matter levels.
- Absence of a healthy microbe community.

Most sources of organic matter contain boron. If a quick boron boost is needed, many gardeners use the borax mined for laundry purposes.

However, keeping existing boron available to plants is also critical. Adjusting pH to a neutral range, increasing the organic matter in the soil, and promoting microbe health are all key steps to take in the long run.

Boron toxicity is not a problem in most parts of the world. It can occur in areas where there is an underground borax deposit in contact with water or volcanic gases. Excess boron can also build up in the soil in arid regions and in fields or gardens irrigated with contaminated water.

The susceptibility of plants to boron toxicity varies widely by species. Some plants, such as azaleas are killed by more than a trace amount of boron. Others, such as cucumbers, seem to be able to absorb nearly any amount of boron with impunity.

Because boron is easily leached from soils, an excess can be flushed from the soil with large amounts of water (assuming your water supply is not the source of the problem). Once this has been completed, it is usually a good idea to check the soil pH and correct as necessary. You can also apply fresh, uncontaminated topsoil in the case of a severe issue.

Calcium

Fast-growing plants are susceptible to calcium deficiency when it is not readily available in a soluble form. Calcium can only move within the plant through the xylem (vascular tissue) as transpiration occurs. Therefore, anything that inhibits the flow of moisture into, through, or out of the plant can lead to calcium deficiency symptoms, including:

- Cold temperatures.
- High humidity.
- Dry soil.
- Excess potassium, phosphorus, magnesium, or sodium.

Also watch out for sandy soils that leach calcium when it rains.

Few soils in nature are truly calcium-deficient. Where this does occur, the following amendments can be used:

- Well water.
- Compost or compost tea.
- Chamomile.
- Epsom salts.
- Seaweed.
- Eggshells.
- Pulverized oyster shells.
- Gypsum.
- Calcitic or dolomitic lime.

But the vast majority of gardeners already have enough calcium to work with; they just need to focus on keeping calcium readily available to the plant. While the weather does not always cooperate, anything you can do to provide a steady supply of moisture to your plants is helpful. Water deeply to encourage root development, and apply mulch to keep the soil moist. Also be careful when hoeing to avoid damaging roots lying close to the surface.

Calcium excess is rare in plants. However, in alkaline soils, an overabundance of calcium can bind to other minerals and prevent plants from absorbing them. Should this occur, correct the soil pH to bring it closer to neutral. Obviously, you will want to stop applying amendments high in calcium, such as dolomite or gypsum, as well.

Chlorine

Chlorine deficiency is not common as far as we know, since chlorine is usually found in both soil and rainwater. However, it can be leached from the soil under wet conditions if the soil contains too much sand and not enough organic matter. Excess nitrate, borate, sulfate, or molybdate can also cause a chlorine deficiency.

Addressing the chlorine deficiency will involve adding organic matter to the soil and correcting any other nutrient imbalances that may exist. You may also need to cut back on the water.

Chlorine toxicity is much more likely to occur than chlorine deficiency. Toxicity can occur naturally in arid regions where there is a buildup of soil salts due to insufficient rainfall. In other parts of the United States, the problem is usually a specific pollution source:

- Chlorinated water.
- Swimming pool runoff.
- Winter road treatments.

Ideally, steps will be taken to identify the source of the chlorine contamination and prevent it from recurring. The garden may need to be relocated or, if this is not possible, moved into containers in a sheltered location. For immediate correction of the problem, flush the chlorine out of the soil with chlorine-free water. Gypsum application may also help.

Cobalt

Soils naturally vary in their cobalt levels. Furthermore, this nutrient is easily depleted in soils that are subject to erosion or leaching.

Even in soils that contain cobalt, signs of deficiency may appear if the pH is unbalanced. The soil pH must be around neutral to enable plants to take up the nutrient efficiently.

Before attempting to correct this problem by applying a fertilizer with cobalt, make sure that the cobalt already in your soil is available to plants. Test the pH and adjust it to around 7.0 if necessary. Also add organic matter for better nutrient retention. If the problem persists, you may need to consider applying a source of cobalt. Manure from animals that received adequate dietary cobalt is a good option. Kelp-based fertilizers are also useful.

Toxic levels of cobalt are rarely found naturally, but can be introduced from outside sources, such as runoff or irrigation water.

If you suspect a cobalt toxicity problem in your garden, you will need to find and address the cause of the contamination. Also make sure that your soil contains an adequate amount of nickel, as that will mitigate the effects of the problem.

Copper

Some soils (usually sandy, peaty, or too acidic) are naturally deficient in copper. Lettuce and onions are among the vegetables most sensitive to this problem, but legumes can also present symptoms.

One way a copper deficiency can be corrected is through the application of copper sulfate. Keep in mind, however, that while copper sulfate is approved for organic use, most of the copper sulfate sold for gardening purposes is manmade. Animal manure and coffee grounds are plant-safe natural sources of copper.

Whatever method you prefer, be sure to correct your pH as needed.

Copper is extremely toxic to plants, so be careful when amending soils with copper or using copper sulfate sprays on plants.

Excess copper tends to stick around in the soil for a long time, as it is not water-soluble. You may be able to mitigate toxicity symptoms by making sure your plants have enough iron. Keep soil pH near neutral and other nutrients balanced. Also take steps to reduce stress on soil microorganisms, such as adding organic matter and reducing or eliminating tillage. In severe cases, you may need to consider moving the garden into raised beds or containers.

Iron

Some soils are naturally iron-deficient, while others are depleted by repeatedly growing iron-hungry plants such as blueberries, rhododendrons, petunias, and snapdragons. In most, cases, however, iron deficiency is brought on by other problems that interfere with a plant's ability to take up the mineral from the soil:

- Soil compaction.
- pH of 8.0 or higher.
- Phosphorus toxicity.

- Calcium toxicity.
- Copper deficiency.
- Excessive soil moisture.
- Root disease.

If you do determine that there is an iron deficiency in your plants, before you run for an iron-rich fertilizer, try to determine the root cause of the problem. Reduce plant stress as much as possible, and work to improve the soil structure and drainage. Also be sure to correct any other nutritional imbalances that emerge. Usually only plants with a particular affinity for iron will need any supplementation.

One excellent source of iron for plants is water that happens to contain it, as this form of the mineral tends to be very easy to absorb. Natural amendments to increase iron levels in the soil include blood meal and composted nettle plants.

Iron toxicity is not common in most soils, but it can occur naturally due to contaminated water or the use of fertilizers containing iron chelates. Plants may also take up too much iron if the pH falls too low or if they are deficient in other nutrients, particularly potassium or zinc.

To correct iron toxicity, first check the pH of your soil. It needs to be at least 5.8 to avoid iron toxicity in sensitive plants. If this fails to solve the problem, you may want to consider more advanced lab testing on your garden soil to identify a nutrient imbalance. Also search for and eliminate any potential sources of contamination.

Magnesium

Soils that are low in magnesium tend to be those that suffer a good deal of nutrient leaching, perhaps due to sandy texture, excessive moisture, or insufficient organic matter. Acidic soils are also prone to magnesium leaching.

In some cases, a deficiency may be induced in plants that have absorbed too much calcium or potassium. Likewise, if plants do not receive enough water, they cannot absorb magnesium from the soil.

When faced with a magnesium deficiency, it is typically best not to supply large applications of high-magnesium fertilizers, as this can create a new mineral imbalance. Instead, use generous amounts of compost to supplement your plants' magnesium take, and then work on improving the overall health of the soil. In particular, correct the pH and amend the soil with organic matter.

If you feel that you must take more drastic measures to raise soil magnesium levels, consider applying Epsom salts or dolomitic lime.

Magnesium toxicity, while fairly rare, can occur naturally when the rock underlying the soil is high in magnesium. It can also be induced by the overuse of dolomitic limestone and similar amendments.

Should this occur in your garden, your best bet is to bring calcium and potassium back into balance. Of course, you will also want to discontinue the use of high-magnesium soil amendments.

Manganese

Even though manganese is common, it can become unavailable due to adverse conditions. Factors that can reduce manganese availability and uptake in plants include:

- Poor soil drainage.
- Soil pH over 7.5.
- Excessively high organic matter levels.
- Insufficient nitrogen.
- Excess iron.

Root crops, particularly onions, have considerable manganese needs. Other plants susceptible to manganese deficiency include peas, green beans, raspberries, apples, and cherries.

This problem can be corrected with manganese chelate. However, it is not safe to apply a manganese fertilizer without first testing the soil to make sure it is truly necessary. Instead, stop applying organic matter for the time being and correct the pH to a range between 5.0 and 7.0. Then apply nitrogen. If iron toxicity is involved, address this problem, as well.

Manganese is toxic to plants when present in large quantities, usually due to fertilizer use. However, manganese toxicity may also occur when the soil pH is below 5.5.

If your plants present signs of manganese toxicity, the first thing to do is to check the pH and bring it up toward neutral, if necessary. Water heavily to flush the excess from the soil, then adjust your garden watering routine to keep the moisture supply as even as possible in the future. Also check your plants for symptoms of other deficiencies and address those as needed.

Molybdenum

Soils that are deficient in molybdenum are very uncommon. Most cases of molybdenum deficiency in plants are caused by low pH. However, molybdenum can be leached from sandy soils that receive too much water. Likewise, peaty soils are commonly low in many nutrients, including molybdenum.

To correct this problem, adjust the pH to near-neutral (at least 6.0). Consider amending sandy soils with organic matter to improve nutrient retention. Peaty soils will probably require supplementation with compost or seaweed extract to boost nutrient levels. If the problem persists, you may want to find an organic seaweed extract fertilizer to use, as seaweed extract usually contains small amounts of molybdenum.

Molybdenum toxicity is nearly unheard of in plants. Plants appear to have an enormous tolerance for this mineral. One thing to watch out for, however, is the use of foliar sprays containing molybdenum, as they can burn the leaves of your plants. Chelated molybdenum is the most hazardous form, but even a natural seaweed spray can cause damage if the concentration is too high.

Nickel

Nickel is fairly common throughout the environment, except in sandy or gravelly soils. However, plants may not be able to absorb enough nickel from the soil if the pH is too high or the soil is contaminated with cadmium. Excess levels of cobalt, copper, iron, magnesium, or zinc can interfere with uptake.

Because application of special high-nickel fertilizers can cause serious toxicity problems, it is usually best to find a fertilizer that offers a good mix of trace elements to correct the problem. Animal manure is a natural option, as are the composted remains of plants that accumulate nickel readily, such as brassicas.

Although nickel toxicity is not common, excessive levels of nickel can occur in soil if too much nickel is applied via fertilizer, manure, or contaminated water.

If your soil does contain too much nickel, identify and eliminate the source of contamination. Also be sure to add plenty of organic matter to reduce the rate at which plants absorb nickel.

Selenium

Selenium levels are usually low in areas where overabundant moisture causes leaching. Most plants do not show deficiency symptoms. Selenium deficiency is more of a concern to the humans who eat the produce.

Should you wish to boost selenium levels in your garden, you can add selenium relatively safely in the form of volcanic ash or saltwater-based fertilizer. The manure from animals fed selenium supplements is another potential source.

Plants need selenium only in very small quantities. Toxicity can readily occur if soils are amended with selenium or if they receive industrial runoff. Some soils naturally contain excess selenium.

To address this problem, you will need to identify and eliminate the source of contamination if possible. If this is not an option, consider switching to container gardening to avoid consuming too much selenium from your produce.

Sodium

Some of the plants Americans enjoy growing in their gardens, such as asparagus and beets, were originally seashore plants that were accustomed to coping with regular salt spray. While these plants appear to adapt to lower sodium levels, they could possibly benefit from a little more sodium than they typically get in an inland garden.

The use of salt in the garden is an interesting topic and is typically reserved for the cultivation of asparagus. Due to asparagus's high tolerance to salt, coarse salt is sometimes used to kill weeds. Asparagus seems to thrive on it.

That said, you probably don't want to run outside with the salt shaker, as most plants have a low tolerance for sodium. In fact, if you use a lot of fertilizer or if you live near a shoreline, your plants could be getting too much sodium. Sodium may also accumulate in soils and groundwater during a drought.

The good news is that excess sodium is easily flushed from the soil with zealous watering from an uncontaminated source. Afterward, you will need to replace any other water-soluble minerals you might have washed away, as well.

Sulfur

Many factors appear to be associated with sulfur deficiency:

- Too much sand or silt.
- Not enough moisture.
- Low soil temperature.
- Low pH.
- Excess phosphorus.
- Low levels of organic matter.

Most forms of organic matter, particularly animal manure, typically contain sulfur and make an excellent natural remedy for this situation. However, you should also try to identify a possible cause of the deficiency so that you can correct the root problem.

Sulfur toxicity is fairly rare now. Toxicity used to result from air pollution due to industrial activities. These days, it shouldn't be a problem unless sulfur-based pesticides are being used in excess. Few soils contain too much sulfur naturally.

Should excess sulfur build up in your soil, water heavily to flush it out.

Zinc

A zinc deficiency is often a sign of soils that have been degraded in some way, usually through erosion or depletion of organic matter. Plants may also suffer from zinc deficiency if their uptake is inhibited by waterlogging, cool soil temperatures, excessively high pH, or high levels of phosphates.

Should a zinc deficiency occur in your soil, correct any drainage issues that need to be addressed and amend the soil with organic matter to improve nutrient retention. Lower the pH to at least 7.5 to improve zinc availability. Also cut back on the level of phosphorus you are adding. Animal manure can be added to boost soil zinc levels once these issues have been addressed.

If your plants have experienced zinc deficiency before, consider planting later to give soil temperatures time to rise in the spring. Alternatively, remove any mulch on the soil temporarily to allow faster warming.

While zinc excess is not common, this mineral is toxic to plants except in small quantities. Deteriorating metal equipment and other forms of pollution can be a problem. Plants may absorb too much zinc when the soil is excessively acidic or when there is excess magnesium in the soil.

Where zinc toxicity has occurred, your best bet is to eliminate any sources of contamination and raise the pH to a near-neutral level.

Striking a Balance

As you may have noticed, a balance between the various minerals is necessary to keep plants from absorbing too much of one thing and too little of something else.

So how do you achieve the right balance? The major key is to feed your soils materials that contain a good mix of nutrients rather than attempting to supplement them with two or three nutrients alone (even NPK). Many commercial fertilizers, even organic fertilizers, focus on NPK to give plants a fast boost, rather than address the slower process of building up both macro- and micronutrients in balance. For this reason, the ideal solution is to gather your own natural materials and apply them as mulch or compost. Keep in mind, however, that homemade compost can still be unbalanced if you use only one or two ingredients and do not seek out a wide range of organic material.

Of course, there are cases where a single nutrient is so depleted through adverse circumstances that it will need to be replaced before any plants can be grown in the garden. These are usually extreme situations, often involving the selection of a less-than-ideal site for the garden (e.g., a former construction zone). In such cases, you may want to consider in-depth soil analysis to determine precisely what and how bad the deficiency is in order to bring nutrient levels back up cost-effectively without creating a new imbalance.

Fertilizers, Natural and Otherwise

For minor nutrient imbalances, it is often best to improve the soil gradually with compost and organic matter, as drastic action can throw the soil into a new imbalance. When a nutrient deficiency is sufficiently severe that plant health and productivity are impaired, however, solutions with more immediate impact may be in order. This is where fertilizers come into play.

Mineral Fertilizers

Mineral fertilizers include rock dusts and powders used to remineralize depleted soil, ranging from granite dust to volcanic basalt. (The limestone used to adjust pH is, of course, a mineral, but because it is not typically used for the purpose of adding nutrients, we won't address it here.) There are also sea minerals, which are really dried sea water.

Advantages of mineral fertilizers include:

- Can be inexpensive, depending on the product used.
- Wide variety of trace minerals, depending on the product used.
- Paramagnetic properties of some minerals.
- Safety for plants (due to their low concentrations, you would be hard pressed to apply too much).

The disadvantages of mineral fertilizers include:

- Comparatively low amounts of nutrients.
- Minerals in a form that cannot be immediately used by plants without further decomposition.

If your plants need help pronto, mineral fertilizers are not going to solve the problem, most likely. Remineralization is carried out for the long-term benefits, not the immediate results.

Furthermore, research suggests that mineral fertilizers are most effective in acidic soils under conditions that promote rapid nutrient leaching. So if you have leached, acidic soil, it may be worth considering rock dust as a way to apply a balanced range of nutrients safely.

Organic Fertilizers

Organic fertilizers are those that are derived from once-living things, whether animal or vegetable. As we are using the term, *organic fertilizer* does not just refer to those expensive solutions you can purchase, but also to things you can find or make yourself, like lawn clippings and compost.

Commercial organic fertilizers tend to approach plant health from the same angle as chemical fertilizers (providing a quick boost of a few particular nutrients in a form readily available to plants). However, they rely on natural ingredients, particularly seaweed and bat guano, rather than synthesized ingredients.

Examples of organic fertilizers you may want to consider include:

- Compost.
- Animal manure.
- Bone meal.

Advantages of organic fertilizers include:

- Virtually free if made at home.
- Less risk of salt build-up than with chemical fertilizers.
- Ability to improve soil structure.
- Low risk of toxicity to plants.
- Generally safer for pets, wildlife, children, and others.

Disadvantages include:

- Variability.
- High cost if purchased.
- Poor nutrient balance if you use only one or two ingredients.
- Slower uptake than with chemical fertilizers.
- Dependence on a healthy microbe community to break them down.

Every garden needs organic fertilizers for peak health. However, it does take time to see the results.

Chemical Fertilizers

Chemical fertilizers may be derived from mineral and organic sources, but they have undergone extensive processing to extract their nutrients.

Chemical fertilizers are made to feed the plant rather than the soil. They provide a brief but intense boost of a few selected nutrients that often produces dramatic growth and, hopefully, increased productivity.

The benefits of chemical fertilizers include:

- Low cost.
- Precise formulation.
- Fast uptake.

The disadvantages of chemical fertilizers include:

- Rapid growth of soft, disease-prone plant tissues.
- Ability to kill plants if too much is applied.
- Gradual accumulation of toxic salts in the soil.
- Disruption of the soil community.

Without a doubt, chemical fertilizer is a quick fix. Not only does it not improve the soil, in the long run it has the potential to harm the soil. Situations where chemical fertilizer may be useful include container gardening.

Chapter 8
Building Organic Matter and Humus

You may have noticed that we have already referred to organic matter and humus a great deal in this book. These two ingredients of healthy soil have many benefits, some of which we have already discussed:

- Putting the proper spacing between soil particles for maintenance of air/water balance.
- Protecting soil nutrients from leaching out during rainfall.
- Feeding the soil community, including earthworms.

In this chapter, we will learn about additional benefits of organic matter and humus, as well as how to set about adding them to garden soils.

What is Organic Matter?

We have already learned a bit about the organic matter component in the soil. You may recall that organic matter is derived from decomposing plant and animal matter. In nature, this includes old leaves, grass residue, old roots, fallen tree limbs, and animal manure.

In the soil, organic matter can be found in various states of decomposition. It may be hardly decomposed at all, and thus be very coarse. Organic matter in this raw state is not yet suitable for garden plant consumption, but it is an excellent feast for microorganisms. Over time, the materials in organic matter will break down and start to release nutrients in a form available to plants.

As microorganisms digest the organic matter, the original substance will become increasingly unrecognizable. Eventually, when the decomposition process is complete, something amazing happens. The organic matter transforms into humus.

What is Humus?

Humus is distinct from organic matter because it has no shape or structure—not even cells. However, it is also derived from organic matter, consisting of fully decomposed plant and/or animal residues.

There are a number of related gardening terms that humus can be confused with:

- **Organic matter.** Organic matter is still in a state of decomposition. Some of the ingredients that went into it can be identified and separated from the rest.
- **Compost.** Compost is closely related to organic matter. It still contains particles of residue that can be further decomposed. Finished compost may present to the naked eye no sign of the organic matter that went into it, but there are still microscopic structures present that are not found in humus.
- **Peat.** Peat is a constant source of confusion for gardeners. Peat moss is not peat humus, and peat humus is not humus at all. Peat moss is partially decayed sphagnum moss, a species of moss that grows in bogs. Peat humus is a combination of sediments and more thoroughly decayed peat moss that collects at the bottoms of bogs.

And as for the bagged humus you picked up at the garden center, that was probably nothing more than compost. A looser definition of the word *humus* is typically used for commercial purposes.

Humus, to put it as simply as possible, is the part of the soil that is without structure and is made up of entirely decomposed plant and animal remains.

Contrary to popular belief, humus has no nutrients, since it is fully decomposed. Instead, structureless humus paradoxically plays more of a structural role in garden soil.

The following are a few of the functions that humus serves:

- Adding to soil density.
- Giving the soil a loamy texture.
- Absorbing heat from sunlight.
- Retaining moisture.
- Retaining the nutrients found in other parts of the soil.

In short, humus takes the things that plants need to live and places them within easy reach of the roots. Scientists believe that humus may also have a hormone effect on plants, regulating their growth and reproduction.

How to Build Organic Matter and Humus

There are three primary ways that gardeners build the organic matter content of their soils:

- Mulching.
- Composting.
- Cover cropping.

Each method has its own pros and cons, so consider implementing a combination of methods.

Mulch

If you are new to gardening, you definitely need to consider mulch. Mulch is good for both you and your plants.

Pros and Cons of Mulch

Mulch keeps weeds in check. It covers up bare soil and keeps weed seeds from germinating. If a weed does manage to sprout, it stands a good chance of being smothered. And as for those few weeds hardy enough to poke their leaves up through the mulch, they will be spindly and rooted in moist, loose soil, and therefore easy to pull.

Mulch improves the soil. Bare soil is typically not healthy. Lying exposed to hot suns, drying winds, and pounding rains is a sure recipe for hardpan. In fact, the weeds that spring up on bare soil are nature's tools for healing it and providing it with a protective cover.

Mulch works to improve the soil in both the short term and the long term. In the short term, it prevents the soil from hardening into a brick, thus providing an immediate improvement in soil structure and aeration. It also moderates the soil temperature, creating a friendlier habitat for garden plants and soil-building earthworms.

In the long term, mulch decomposes and adds vital nutrients and organic matter to the soil. It is staggering how much healthy soil can be built over one gardening season just by using mulch and compost. As your soil grows and improves, your plants will become healthier and more vibrant, better able to ward off insect pests.

Mulch keeps soil moisture steady. In wet weather, mulch is useful in preventing your plants from drowning. As rain falls, the mulch intercepts the drops, preventing them from compacting the soil and forcing them to trickle down slowly to root level. In the process, the mulch itself will absorb some of the excess moisture. The organic matter added to the soil by decomposing mulch will also help out by allowing any surplus rainfall to drain away from the roots, ensuring that the plant receives an supply of adequate air.

In dry weather, mulch is a must because of its water-conserving properties. Mulch protects the soil from rapid drying due to sun and wind. Without mulch, you may have to water your entire garden every day in the summer. With mulch, you can water less frequently, promoting deeper root growth that will in turn make your plants more drought-hardy.

Are there any disadvantages to mulching? Actually, there are. Here are a few pitfalls you need to be aware of:

- Mulch slows down soil warm-up in the spring.
- Mulch may pull nitrogen out of the soil as it decomposes.
- In wet weather, mulch can harbor slugs and similar pests.
- Some mulches contain undesirable elements, such as herbicide residue or weed seeds.

However, for every problem, there is a solution:

- Pull aside the mulch temporarily early in the spring to expose the soil to the sun.
- Choose a mulch that contains its own nitrogen, or allow the mulch to partially decompose before applying it.
- Apply diatomaceous earth to your mulch as a natural pesticide, or choose a mulch that slugs don't like, such as seaweed, raw wool, or pine needles.
- Source your mulch carefully to avoid contamination.

Choosing the Right Mulch

If you have decided to mulch your garden, your next consideration will be choosing the best mulch to use. First off, it is important to recognize that everyone's circumstances are unique. Factors that will affect your choice of a mulch include:

- The types of plants you will mulch.
- The materials available in your area.
- The cost of the different local materials.
- Your purpose for mulching (weed control, building organic matter, conserving moisture, etc.).

Mulches can be divided into two basic categories:

- Organic (derived from plant or animal matter).
- Inorganic (derived from rock or manmade materials).

Inorganic mulches, such as gravel or plastic, avoid some of the difficulties commonly associated with mulch. However, they do not contribute organic matter to the soil and therefore will not be discussed here. Suffice it to say that because they do not break down they are generally most effective around perennial plants and can be inconvenient to work with in a vegetable garden, where you must frequently access the soil to plant seeds, transplant seedlings, and harvest root crops.

When it comes to organic mulches, you have several options, each with its own pros and cons.

Straw

Straw and hay are not the same. Straw is simply crop stubble, while hay is the entire grass plant, including seed heads. Straw is an effective weed barrier, while hay is primarily a simple mechanism for introducing new weeds to your garden. Make sure you mulch with straw, not hay.

The pros of using straw are:

- Availability.
- Low cost.
- Generally attractive appearance.
- Effectiveness as a weed barrier if applied thickly enough.
- Ability to protect plants from frost.
- Cooling to the soil in hot weather.

- Soil structure improvements.
- Excellent moisture retention.
- Prevention of soil erosion.
- Slow decomposition.

The cons of straw include:

- Tendency to attract rodents.
- Tendency to attract slugs in cool, wet climates.
- Possibility of weed seeds.
- Possibility of mold.

Straw often works best for general-purpose vegetable garden mulching, summer through winter.

Lawn Clippings

Lawn clippings should be used with care. They should be applied only when dry and preferably in conjunction with coarser materials to avoid forming a heavy, moldy, anaerobic mat. Lawn clippings from yards that contain a dog or cat are not safe for use on food crops because pet urine and feces can contain bacteria that are very dangerous to humans. Also, if accepting bagged lawn clippings from other people, always check to make sure that they do not use herbicides, as accidental contamination can spell speedy death to your flowers and vegetables.

The pros of lawn clippings are:

- Availability at no cost.
- Added nitrogen.
- Excellent weed control provided the clippings do not contain weed seeds.
- Soil moisture conservation.

The cons of lawn clippings are:

- Possibility of contamination with pet waste.
- Possibility of contamination with herbicides and other chemicals.
- Possibility of weed seeds.
- Mold if applied too thickly or when damp.
- Odor if applied too thickly or when damp.
- Tendency to form a water-repellent mat if applied when damp.
- Extremely fast decomposition.

Lawn clippings often work best for mulching summer vegetables with high nitrogen requirements.

Leaves

Leaves can improve your soil structure and nutrient profile in an amazingly short amount of time. Unfortunately, they can also be difficult to manage in the garden, as they blow away when dry and form heavy, waterproof mats when wet. Want to avoid some of the problems associated with using leaves as mulch? Shred or partially compost the leaves before applying them. A final warning—do not use walnut leaves, as they are toxic to many plants.

The pros of leaves as mulch include:

- Availability at no cost.
- Ability to protect plants from frost.
- Ability to cool the soil in summer.
- Good soil moisture retention.
- Improvement in soil structure.
- Wide range of nutrients.
- Earthworm health.

The cons of leaves as mulch include:

- Tendency to blow away.
- Plant diseases, depending on the source.

- Formation of a water-repellent mat after rainfall.
- Tendency to deplete soil nitrogen in the short term unless partially composted.

The best application for leaves is protecting and enriching vegetable garden soil over the winter.

Pine Straw

Pine straw is simply pine needles used as mulch.

The advantages of pine straw or needles as mulch include:

- Low (or no) cost.
- Attractive appearance.
- Effective weed control.
- Ability to repel slugs to some degree.
- Improved soil structure and drainage.
- Prevention of soil erosion.
- Water penetration.
- Slow decomposition.

The cons of pine straw or needles include:

- Sharp points that are painful to handle—wear gloves.
- pH too low for many plants.
- Inhibited earthworm activity.

Pine straw or needles are best for mulching acid-loving plants such as blueberries, hydrangeas, or rhododendrons.

Wood Mulch

There are many varieties of wood mulch to choose from, all with their own unique benefits. Cedar even provides a certain level of protection from insects. Just watch out for toxins—if you purchase your wood mulch, stay away from mulches containing dyes. Also, never use black walnut, as it contains toxins lethal to many garden plants.

The advantages of using wood chips or bark as mulch include:

- Availability, sometimes for free.
- Attractive appearance.
- Effective weed control.
- Fairly good retention of existing soil nutrients.
- Long-term boost in soil nutrients.
- Slow decomposition.

The disadvantages of wood-based mulches include:

- Dyes.
- Tendency to float away during heavy rainfalls.
- Depletion of soil nitrogen in the short term.

Wood mulches are frequently best as a ground cover for perennial beds.

Compost

Compost can be used as a mulch, too (we'll delve into how to make it in the next section).

The pros of using compost for mulch are:

- Availability at no cost.
- Considerable improvement in soil texture and structure.
- A boost in nutrients.
- Promotion of a healthy soil microbe community.

The cons of using compost for mulch are:

- Possibility of spreading plant diseases.
- Only partial weed control.
- Fast decomposition.

Typically, the best way to use a compost mulch is to spread it thinly underneath another mulch material to feed garden plants of all types.

Peat Moss

Peat or sphagnum moss is harvested from peat bogs. It has unusual characteristics that can be either good or bad depending on your requirements. For one thing, it can absorb water like a sponge, which improves boggy soil considerably but may allow plants to dry out in hot, droughty weather. For another thing, it makes the soil more acidic, although typically not enough to present a problem.

The advantages of using peat moss for mulch include:

- Availability.
- Considerable improvement in soil structure and drainage.
- Slow decomposition.

The disadvantages of using peat moss for mulch include:

- Susceptibility to blowing away if dry.
- No nutrients.
- Tendency to hinder water penetration if applied too thickly.

Peat moss mulch often works best when improving heavy soils that drain poorly.

Sawdust

If you do plenty of woodworking, sawdust is definitely a mulch you should consider. But be careful about the sawdust you use—dust from treated lumber can add toxins to your soil.

The advantages of sawdust for mulch include:

- Availability at no cost.
- Ability to control weeds well if applied thickly enough.

The disadvantages of sawdust for mulch include:

- Formation of a water-repellent crust after rain.
- pH too low for some plants.

- Depletion of soil nitrogen levels.
- Rapid decomposition.

Sawdust is a good way to mulch acid-loving plants inexpensively.

Cardboard

Cardboard is the go-to mulch if you have a serious weed problem—nothing can penetrate it! Keep in mind, though, that cardboard can be difficult to deal with. You will want large sheets to cover as much surface area as possible without leaving cracks for weeds to grow through, and you will want to cover it with straw or another attractive mulch to keep it in place but out of sight.

The advantages of cardboard mulch are:

- Availability, often at no cost.
- Extremely effective weed control.

The disadvantages of cardboard mulch are:

- Extremely unattractive appearance unless completely covered by another mulch material.
- Difficulty of planting or transplanting in gardens that have been mulched with cardboard.
- Boron toxicity due to glue; before using, soak in water until the cardboard is soft, then discard the water (or use it as a fertilizer for strawberries).

Cardboard is great for putting the brakes on heavy weed infestations in perennial beds when used in combination with another mulch material.

Newspaper

While many gardeners avoid mulching with newspaper for fear of lead contamination, newspapers phased out lead-based inks long ago. Black-and-white newsprint is perfectly safe for the garden these days; colored inks may still contain some heavy metals. Note that newspaper is prone to blowing around. Cover it with another mulch, such as straw or wood chips, to avoid inadvertently trashing the neighborhood.

The pros of newspaper are:

- Availability.
- Ability to cool the soil in summer.
- Effective weed control.

The cons of newspaper are:

- Tendency to blow away.
- Unattractive appearance unless thoroughly covered.
- Heavy metals if colored ink was used.

Newspaper is ideal for preventing weed growth between rows in a vegetable garden when combined with straw or another mulch material.

A Final Note

Whatever mulch you choose, apply it thickly. A dense mulch such as wood chips can be spread on six inches deep; a light, airy mulch such as dry straw will need to be a foot deep to be effective.

Compost

Compost is beloved by all who garden organically, and little wonder. Applying compost to your garden offers fast, obvious improvements in soil structure and vitality.

Unfortunately, a certain mystique has come to surround compost-making, leaving beginning gardeners with the impression that process is touchy and difficult. Actually, there is more than one method of making compost, each with its own pros and cons.

Hot Composting

Most gardening books focus primarily (sometimes exclusively) on hot composting methods.

The key to successful hot composting is achieving the right balance of ingredients. Every hot compost pile needs materials high in nitrogen (called "green materials") and materials high in carbon (called "brown materials"). The green materials feed the microbes that will do the work, thereby fueling the whole process. The brown materials add bulk and air flow, preventing anaerobic decay from taking over and creating a foul smell. Experts disagree on the proper balance between browns and greens, but a ratio of two browns to one green or even one to one has consistently been proven effective for most home gardeners. For faster, hotter decomposition, the ingredients should be chopped small. Layering greens and browns is not essential.

Most gardeners also like to add an activator to the pile to ensure that the necessary organisms are present and ready for work. Many expensive preparations are available for purchase, but good-quality soil is quite sufficient for the purpose. Yarrow leaves are also beneficial, as they add minerals that can help the process go quickly and efficiently.

The compost pile should be a three-foot cube to allow the center to hold heat efficiently. A four-foot cube is better, while a five-foot-cube is about the maximum size for effective composting.

Once the ingredients are properly balanced, special microorganisms take over the pile and begin the decomposition process, emitting heat as they work. The heat builds up quickly in a well-made compost pile, ideally reaching temperatures between 120°F and 150°F.

From time to time, the microbes will use up the oxygen in the center of the pile, causing the temperature to drop. For this reason, the pile needs to be turned once a week or more, giving the tiny creatures access to fresh air and bringing organic matter from the outside of the pile into the center for processing.

At the same time, the moisture content of the pile should be considered. The pile should be moist, but not soggy. In dry weather, you will probably have to spray the pile down with a hose while turning it. In wet weather, you may want to place a tarp over the pile or make your compost in a fully contained bin to avoid drowning the microorganisms.

After about a month, the compost should look dark and crumbly (think chocolate cake crumbs), with no recognizable remnants of the former ingredients. At this point, the pile will no longer heat up when turned. The compost is technically finished, but it will benefit from one or two weeks of curing. This allows beneficial organisms a chance to repopulate the compost.

The benefits of hot composting are:

- **Speed.** Hot compost does its thing in short order, making the finished product available within three to four weeks.
- **Reduced space needs.** A hot compost pile can be built when you need it and removed in as little as a month. In contrast, a cold compost pile can take up garden space for as long as a year, requiring a commitment in advance.
- **Full control.** Some gardeners like the control hot composting gives them over the finished product. They can easily adjust the temperature or decomposition speed by turning the pile. To them, monitoring the pile is not a chore; it is an art.
- **Sterilization.** This is probably the #1 reason that gardeners choose hot composting methods. As the compost pile heats up, many pathogens and weed seeds get fried.
- **Toxin degradation.** Are you using ingredients that were not grown organically? Hot composting can break down and neutralize the pesticides present on that orange peel and the residue of any medications you might have administered to your horse that are now lurking in the stable bedding.
- **No pests.** Assuming the composting process is working correctly, that is. Most insects and rodents will leave a hot compost pile

alone. A poorly made compost pile can still attract every type of pest from ants to raccoons.

- **Helpful heat.** The heat emitted by a good compost pile can be put to work. One of the most common uses for this composting byproduct is heating a greenhouse.

The disadvantages of hot composting are:

- **Difficulty.** After unsuccessfully battling a compost pile that simply refuses to heat up or that smells and draws flies, many gardeners are tempted to give up on composting altogether. Hot composting is part art, part science. It takes practice. (Here's a hint—make your compost in summer, when naturally hot temperatures will give you a head start.)
- **Need for ample resources.** A hot compost pile cannot be built in stages. You must have the necessary ingredients on hand and in the right balance on day one.
- **Time and effort.** A hot compost pile needs frequent supervision. Besides weekly turning and watering, it will need to be monitored to ensure that it is not too hot or too cold.
- **Lack of natural equivalents.** Most gardeners who use compost are already focused on growing healthy plants naturally. One question that comes up periodically is why hot composting is so widely advocated if there are no equivalents in nature. It appears that most of the composting done in the wild uses the cold method. There may be benefits to cold composting that are lost with hot composting.
- **Microbe and worm death.** Any small organisms that don't like heat will die or escape the compost pile as it heats up, including worms, beneficial fungi, and some lower-temperature bacteria necessary to fight garden disease. Proper curing can remedy this, but it is important to recognize that hot compost still tends to be biased toward heat-friendly bacteria instead of the diverse, fungus-based biology commonly found in more natural

decomposition processes. Some composters feel that bacterial dominance makes the finished compost better suited to grassland plants than to woodland plants. If the compost pile gets too hot (over 150°F) for more than a couple of hours, the problem is compounded significantly.

- **Nutrient loss.** Valuable nutrients are definitely lost if the compost pile overheats. If you smell a strong odor or see vapor rising from your pile, you can assume that nutrients are escaping into the atmosphere. Some gardeners feel that a certain degree of nutrient loss occurs in hot composting even if the temperatures stay below 150°F.
- **Fire hazard.** While not a common occurrence, spontaneous compost pile combustion can and occasionally does happen, for the same reason that wet hay bales sometimes burst into flames. Avoid this problem by monitoring the temperature and keeping it below 150°F; making sure there are no pockets of dry material helps, as well. Also, if using wood ashes, make sure they are completely cool before adding them to the pile.

There are two main reasons that gardeners prefer hot composting:

- To destroy weed seeds and pathogens.
- To create finished compost quickly.

In exchange for these two primary benefits, the gardener must devote considerable planning, attention, and effort to the project. For a busy gardener, this is sufficient reason to turn to cold composting.

If, however, you have the time and resources to have a go at it, you may find hot composting to be rewarding and effective, especially for disposing of materials with potential chemical, disease, or weed seed residues. However, don't let your pile get over 150°F, and allow it to cure for at least a week before using. This will help ensure that the finished compost retains as many of the nutrients and organisms necessary for plant health as possible.

Cold Composting

Are you overwhelmed just thinking about all that preparation and turning? Don't give up on composting altogether yet. First consider cold composting.

A cold compost pile can be started with whatever materials are on hand. The pile can be built gradually over time.

While a precise balance of "greens" (organic matter high in nitrogen) and "browns" (organic matter high in carbon) is not needed for success, incorporating both into the pile will help the decomposition process go faster. Likewise, large pieces of material, such as sticks, can be chopped for faster composting, but this is not necessary.

As with hot composting, air is a requirement to avoid the accumulation of smelly anaerobic bacteria. Air can be introduced by periodic turning, but again this is not essential. One way to allow air circulation within the pile is to pay attention to how ingredients are layered. For example, some coarse (but not too long) sticks in the center will help. Also, soggy materials that stick together like fresh grass clippings or wet leaves should be allowed to dry slightly before going into the pile. If for any reason air flow becomes a problem during the composting process, air holes can be driven into the pile with a stake or piece of rebar.

As decomposition progresses, the moisture content should be monitored. If persistent rains are a problem, the pile should be covered with a tarp to avoid drowning and subsequent anaerobic decay. In dry weather, periodic sprinkling with a hose will be beneficial.

Cold composting goes slowly. How slowly depends on the size of the ingredients, the ratio of greens to browns, and the frequency with which the pile is turned. These variables also affect the temperature that the pile will reach. A maintenance-free cold pile may never hit 90°F. Piles that contain green material and are turned periodically may get somewhat warmer.

The bottom of the compost pile decays first, usually leaving large pieces of material intact on top.

The benefits of cold composting are:

- **Simplicity.** Getting a compost pile to heat up is not easy. Conditions have to be about perfect to make it happen. If gardening is not your full-time job, you may appreciate the ease with which a cold compost pile can be built and maintained.
- **Flexibility.** One of the biggest advantages of cold composting is that you can adapt it to your needs. Don't have much to compost? Toss on what you do have and wait for another opportunity. Don't have time to turn the pile today? It'll wait until tomorrow, or next week, or next month. Ready to cut the process short? Throw away the larger uncomposted pieces of material and dig out the finished compost underneath to use immediately.
- **Fewer material requirements.** With hot composting, you have to have three, four, or five cubic feet of organic matter (properly balanced between greens and browns) ready to go at once. With cold composting, you can add materials as you accumulate them. One day, you can add the grass clippings. A few days later, you can throw a banana peel on top.
- **Compost year-round.** Many gardeners become frustrated after trying to get a compost pile to heat up in cool weather. With cold composting, you can let the decomposition process continue all year. Even in January, freeze-thaw cycles and periodic introduction of moisture in the form of snow will help mechanically break down materials, long before the microbes awake from their winter slumber.
- **Minimal turning.** Let's face it. Not every gardener can find time to turn a bulky compost pile every few days. With cold composting, you can simply turn the pile whenever you have a spare half hour or so. Or you can let it go from start to finish

without ever picking up a shovel. This also makes cold composting a great option for an older gardener who wants to garden naturally but finds hot composting backbreaking.

- **Nature's way.** Where in nature have you seen a hot compost pile steaming away, with obliging creatures turning it at just the right time? Probably nowhere. Nature makes compost slowly.
- **Beneficial organisms.** Many microscopic forms of life can survive in the lower temperatures of a cold compost pile. This includes beneficial fungi and bacteria that thrive on moderate temperatures. These are exactly the microorganisms that your plants need to stay healthy and fight disease. Some gardeners speculate that the preponderance of fungi in cold compost is just right for raising trees.

The disadvantages of cold composting are:

- **The ugly factor.** As much as natural gardeners love good soil, a half-rotted compost pile may not be something they want to see out their kitchen window every day. While a hot compost pile can do its work and be hauled away in a few weeks, a cold pile will likely be sitting in the same place decaying all summer long. Choose your location accordingly.
- **Space requirements.** A cold compost pile will be taking up space for many months at a time. When choosing a place to build a pile, find a corner of the garden that is out of the way, but not too far from the beds to be convenient.
- **Slow going.** Cold composting takes about six months to a year—sometimes longer. While you can speed up the process by building the pile all at once or by turning it more frequently, if you need compost fast, you definitely need to consider hot composting.
- **Anaerobic decomposition.** Cold piles can collapse over time and lose their ability to circulate air, especially if they contain ingredients that tend to stick together like wet grass or leaves. A

smelly bacterial mess that draws flies can be the result. With a hot pile, the frequent turning is sufficient to introduce air into the center of the pile. Cold compost piles should be aerated from time to time; of course, this means that they are not exactly maintenance-free.

- **Pest attraction.** Bugs and mammals love cold compost piles. Chances are you are going to see a lot of flies and rodents. Fencing can help keep away larger pests, such as raccoons and stray dogs.
- **Chemical persistence.** Heat is necessary to break down any chemical residue, such as pesticides and herbicides, remaining on compost ingredients. If you use chemicals on your lawn, do not add the grass clippings to a cold compost pile.
- **Weed survival.** Weed seeds will make it through the low temperatures of cold composting intact, guaranteed. If you composted pulled weeds, you can expect them to put in a second appearance after you apply the compost. For best results, never compost invasive weeds, or noninvasive weeds that have gone to seed.
- **Disease survival.** Most diseases can also survive the cold composting process. If you will not be maintaining a hot pile, you should definitely avoid any diseased plant material. How you dispose of diseased plants will depend on what is legal in your area. Burning them or shipping them out with your trash is ideal.
- **Nutrient leaching.** Depending on how long the composting process is drawn out and on the weather conditions at the time, nutrients can leach out of the pile with every rainfall. Using a composting container will help contain the valuable nutrients your plants need. Covering the pile with a tarp in wet weather is another possibility. Also, you can periodically dig out the finished compost from the bottom of the pile to use immediately.
- **Uneven decomposition.** Remember, the compost will decay from the bottom up. While you're waiting and waiting for the organic matter on top of the pile to break down, there is good compost

that you could be using accumulating underneath. Even when finished, there will likely be large particles remaining, resulting in a coarser final product than is achieved with hot composting.

The reason that busy gardeners often choose cold composting should be obvious by now—a cold pile can be as low-maintenance as desired.

While this is an important consideration, there is another factor that should be given some thought, and this is the quality of the finished product. Cold composting is definitely a more natural method of composting than hot composting, and it definitely fosters more of the beneficial organisms necessary for plant health. Under optimal conditions (i.e., if not dragged out too long or flooded by repeated rainfalls), it may contain more nutrients.

However, there are certain circumstances under which cold compost will be of doubtful quality. If any of the ingredients contain chemicals, disease organisms, or weed seeds, these contaminants will be perpetuated in the compost. Therefore, gardeners who have doubts about the ingredients should either discard them or use the hot composting method.

Finally, time is also a factor. For those who don't mind waiting several months for finished compost, cold composting is a viable option. Gardeners who need to build and amend the soil quickly will benefit from the hot method.

Vermicomposting

Vermicomposting is the process of creating compost with the aid of earthworms. While all composting relies on microbes to do much of the work, vermicomposting allows worms to come to their assistance.

The worm starts the composting process by ingesting organic matter and breaking it down with digestive enzymes. However, the digestive system of an earthworm only absorbs about 5% to 10% of the food the worm eats. This is good news for vermicomposting, because the rest of the organic matter, now moistened and considerably broken down, is excreted and offers a rich buffet for hungry microbes.

Note that vermicompost and worm castings are not the same. Worm castings are simply earthworm excrement. Vermicompost includes worm castings, but also worm bedding, food, and remains in various stages of decomposition, all supporting a vibrant microbe community.

Vermicompost is typically made in a "worm bin" filled with bedding (usually shredded paper) and organic material for the worms to eat. A suitable worm habitat is further created by keeping the bin dark and moist.

Two worm species are typically used:

- Red wiggler (*Eisensia fetida*).
- Red worm (*Lumbricus rebellus*).

The red wiggler is particularly popular because it is easy to care for and produces castings quickly and efficiently.

Of course, you can theoretically dig up any old earthworms in your backyard to populate the worm bin, but there is no guarantee that you will find effective compost-building species this way. A surefire solution is to order red wigglers online.

Finished vermicompost looks much like high-quality soil, but it is incredibly rich in microbes.

Most kitchen scraps and waste paper products make good worm food. However, you will want to avoid anything that produces a foul odor, especially if your worm bins are indoors. This includes meats, oils, dairy products, onions, garlic, potatoes, and anything in the mustard family, such as broccoli. Also note that worms are not a replacement for the neighborhood recycling facility—they cannot process plastic or aluminum.

Vermicomposting offers the following benefits:

- **Quick and easy.** Vermicomposting does not require the precision or labor that hot composting requires to turn out a successful batch. And it has a distinct advantage over cold composting methods—it's quick!
- **Year-round composting.** Worm bins can be kept inside in the winter, allowing you to make compost for use first thing in the spring.
- **High in beneficial soil microbes.** Microbes thrive where worms work, and they are particularly abundant in worm castings. Adding vermicompost to depleted soils can yield dramatic results in garden health.
- **All the benefits of organic matter.** Soil that has been amended with vermicompost will not dry out quite so quickly in the hot summer sun, and it has improved texture and nutrient content. (Note that this is true of all forms of composting.)
- **A good use for kitchen scraps.** Well-fed worms will repay you by improving your garden soil, which in turn will bring more (and healthier) produce into your kitchen.

There are some drawbacks:

- **Expense.** Remember, you will only have reliable results with certain worm species. This means you will likely have to buy worms, adding to your gardening costs. (You will also need to purchase a bin the first time out.)
- **Small quantities.** Regular composting methods can produce more compost much quicker than earthworms can. You will want to use your vermicompost selectively to feed the plants that need it the most.

Vermicompost is equally suited to trees, vegetables, flowers, and potted plants.

Because vermicompost is made in relatively small quantities, gardeners will want to use it wisely. A good rule of thumb is to use it to fuel rapidly growing plants.

The best place to apply vermicompost is around the plant's drip line. Imagine a circle on the ground around the plant roughly marking its circumference. This circle is called the drip line because, if you were to spray the plant with water, this is where the water would drip from the leaves. Many hungry roots are waiting just below the drip line. Spread the vermicompost on top of the soil along the drip line.

So will vermicomposting meet your needs? Likely not if you have a large garden. For a small garden (or for a good hands-on science project), however, it can be a quick way to improve poor soil.

Cover Crops

Cover crops can offer numerous benefits to vegetable gardens:

- Rapid building of organic matter.
- Improving soil drainage.
- Smothering weeds.
- Serving as a trap crop for insect pests (a trap crop is a crop planted for the purpose of attracting pests away from more valuable plants; once the trap crop is infested, it is destroyed, pests and all).

However, care must be taken to choose the right cover crop for the application, as some cover crops will inhibit plant growth.

A few common garden cover crops and their applications include:

- **Annual ryegrass.** Weed suppression. Winter-hardy. Hard to cut down by hand.
- **Buckwheat.** Weed suppression, pollinator attraction. Fast-growing, quick to decompose. Cold-sensitive.

- **Crimson clover.** Adding nitrogen, attracting earthworms. Suitability for heavy soils. Low germination rates.
- **Hairy vetch.** Preventing erosion, adding nitrogen. Suited to many soil types. Spreads aggressively.
- **Winter rye.** Preventing erosion, weed suppression. Suited to harsh winters, fast-growing. Slow to decompose.

For maximum benefits, you might consider sowing both a grass and a legume.

Growing a cover crop is easy:

1. Prepare a weed-free seedbed, just like you would for any other crop.
2. Scatter the seeds evenly over the soil.
3. Rake the seeds in.
4. Water thoroughly.
5. Keep watering often enough to keep the soil moist until the crop germinates and becomes established.

When it comes time to terminate the cover crop, you have several options:

- Till it in.
- Fork it in.
- Mow it down.
- Let it die naturally (be aware that it might come back from seed if you do this).

Can You Have Too Much Organic Matter?

Yes, the mineral element needs to be present for plants to germinate and thrive. They do not do well in soils too high in organic matter (like peat bogs). Ideally, organic matter will compose about 5% of the soil content.

Chapter 9
Building a Soil Community

By now, you've learned a great deal about soil chemistry. However, chemistry is pointless without biology. After all, the point of gardening is to bring plants to life so that they can in turn sustain human life. Plant and human life is ultimately sustained by soil life.

Meet Your Local Soil Residents

Soil residents include the following:

- Fungi.
- Algae.
- Bacteria.
- Protozoa.
- Mollusks.
- Nematodes.
- Earthworms.
- Arthropods.
- Insects.
- Reptiles.
- Rodents.
- Larger burrowing mammals.

Different soils have different proportions of these organisms.

Inoculants

If you are concerned that you don't have a soil community to start with, you may consider using inoculants, amendments that introduce new life to the soil.

Purchased inoculants are frequently used with legumes. The seeds are coated in a solution containing beneficial bacteria to help the growing plant fix nitrogen. You can also purchase inoculants specifically blended for use in the compost pile.

Inoculants do not have to be purchased amendments, however. Undisturbed soil is a surprisingly effective inoculant. If you have access to such soil, dig up a trowelful and place it in a bucket, then roll your seeds in it to cover them thoroughly (this is easier with large, easy-to-handle seeds).

Another option is to bring in animal manure from a trusted source. Avoid manure from animals that are unhealthy or given antibiotics, wormers, or other chemicals, as this manure will not contain the desired microbes. Healthy manure from poultry or ruminants, however, will introduce a vibrant community to your soil. To use manure safely and avoid the risk of contaminating your produce, either add it to your compost pile or give it about six months to age before applying. Also, wait at least 120 days after application to harvest any produce (when growing vegetables, you may want to apply manure in late fall or early winter after harvest).

Feed the Soil, Not the Plants

This phrase is almost a cliché in the organic gardening world, but it's worth repeating because it is so important. Too often people apply quick-fix fertilizers to feed their plants but neglect to feed the soil that is essential for long-term plant health and growth. There is sometimes overlap between feeding plants and feeding soil, but the focus must be on the soil.

The good news is that by now you already know a great deal about how to feed the soil. The single biggest thing you can do to feed your soil residents is to keep your soil supplied with organic matter.

The best type of organic matter you can use to feed the soil is compost. Cold compost and vermicompost are ideal for this, because they already contain a healthy community of desirable residents, but hot compost that has been allowed to cure works, too. Compost is by its very nature food for soil microbes, because it would not exist without them. Thus, by spreading compost on your soil regularly, you are introducing both microbes and the food they need to thrive.

Mulch is also a good choice for feeding the soil community. Not only does it add organic matter as it decomposes, it shields the soil from the sun and creates a cool, moist environment in which microbes, earthworms, and other soil life forms can comfortably work.

That said, all types of organic matter are appreciated by the soil community, so any type of organic matter or organic fertilizer that you use is beneficial.

Threats to the Soil Community

Once you have an established soil community, besides feeding it, your next responsibility becomes to protect it from harm.

Chemicals

Chemicals, whether they are pesticides or fertilizers, are commonly believed to disrupt the soil community and create a wide variety of imbalances in the soil.

When it comes to chemical fertilizers, there is much we don't know about their interactions with soil life at present. The research that exists to date suggests that the number of bacteria does not decrease with fertilizer use (assuming the fertilizer is used correctly). However, such fertilizers do make the following changes to the soil:

- Reduce pH.
- Raise nitrogen levels sharply.
- Increase the rate of organic matter decomposition dramatically.

It is therefore conceivable that such fertilizers could promote the growth of some species at the expense of others. Also, some believe that in the long run chemical fertilizers could cause microbe death due to the acceleration of organic matter decomposition, which would ultimately result in starvation of the community if the organic matter were not replenished.

More certain is the detrimental effect of pesticides and fungicides on soil life. These types of chemicals are created to kill, and they must do it effectively or no one would purchase them. While soil life is amazingly resilient and can usually bounce back after one or two applications, repeated use of pesticides and fungicides will eventually cause toxins to build up in the soil and kill off microbes, fungi, and other good guys.

For this reason, those who are interested in restoring soil health will seriously want to consider growing organically, at least as much as possible. The good news is that, as the soil health improves, plant health will improve, too, making these chemicals unnecessary anyway.

However, if you have been using chemicals until now, be prepared for a transition period. It takes time to restore the balance in the soil, and the process is not always smooth. Be proactive in your soil improvement program, but also be prepared to deal with a number of pest and disease issues for several years (probably three to ten depending on how aggressive you are in your soil healing program). While your goal should ultimately be to phase out all pesticides, for a time you may want to consider switching to natural and organic pesticides, such as diatomaceous earth for insects and baking soda spray for fungal diseases, to ease the transition.

Tillage

Are you looking for new ways to improve your garden soil faster? Have you thought about ditching the rototiller?

No-dig gardening, no-till farming's little brother, offers an exciting way to improve soil with less back-breaking labor. It also presents a far more natural way to garden—after all, Nature doesn't own too many rototillers. Every time the soil is inverted the soil community is disrupted, while leaving it in place creates shelter for your friendly microbial helpers.

Are there pitfalls? The answer is yes. Even so, no-dig gardening may be right for your garden.

Let's look at the pros and cons to determine the situations where no-dig gardening will be most effective.

The pros of no-dig gardening include:

- **Easier on the back.** Digging and tilling are hard work. Eliminating those two steps is a great choice for gardeners who are elderly, have back problems, or struggle with lack of time.
- **No damage done to soil life.** Rototillers disrupt the lives of beneficial bacteria, fungi, and earthworms as they work. While these soil communities will recover before the season ends in organic gardens, the traumatic event is a setback to their work. No-dig gardening fosters life in the soil without interruption.
- **Reduced soil compaction.** This one may come as a surprise to you. After all, tillage is supposed to be the way to loosen up the soil in the spring. However, at the farthest depth that the blades can reach, they actually stop turning the soil and start packing it

down. If tillage continues at the same depth every year, the soil immediately below the tillage zone turns into hardpan. No-dig gardens will not suffer from compaction as long as soil health is properly attended to and clearly defined footpaths are provided to avoid trampling the soil.

- **Soil aeration.** No-dig gardening is precisely how plants thrive in nature. No wide-scale tillage and subsequent exposure of bare dirt occurs. Organic matter such as fallen leaves and dead grass lies on top of the soil and decomposes over time. Why is having the organic material on the surface important? Because it encourages earthworms to come up from beneath to reach the material, creating worm tunnels that promote soil aeration.
- **Better use of rainfall.** Tillage leaves large areas of bare soil exposed to the sun, wind, and driving rain. The combined action of baking, evaporation, and pounding results in soil with a hard crust on top. This crust in turn prevents rainfall from readily soaking down to the roots. Soil that has not been tilled stays friable, and the layer of mulch on top further aids in capturing moisture.
- **Fewer weeds.** Tillage brings weed seeds up to the surface of the ground to germinate. This can be used to advantage by tilling repeatedly during the same season, allowing several weed crops to germinate only to meet their demise at the hands of the rotating blades. Unfortunately, a lapse in the tillage routine can create a disaster. No-dig gardening keeps piling layers of soil and mulch on top of the weed seed bank, preventing it from ever sprouting and smothering the few weeds persistent enough to attempt germination. Unless you are introducing new seeds from an outside source, you will end up with fewer and fewer weeds the longer you no-dig.
- **Fewer pests.** Another counterintuitive effect of no-dig gardening. Tillage is often recommended to expose insect eggs and larvae to the elements (and any helpful chickens that happen to be around). But many gardeners feel that, when done with an eye

toward soil health, no-dig gardening seems to attract fewer of the bad bugs, probably because the plants are far healthier overall and have more stable connections with soil life forms that offer protection from attack. (The exception is slugs, as we will see in a moment.)

The cons of no-dig gardening include:

- **Scale limitations.** No-dig gardening implies using mulch and compost. Unless you have the ability to produce industrial quantities of these two ingredients, no-dig gardening will be nearly impossible in a large garden.
- **Difficulties with starting in seriously compacted soil.** For example, former driveways. Ideally, you will choose a different site for your garden and avoid this problem altogether. If this is not an option, you will probably need to till at least once, likely more than once. The good news is that in this case you will not disrupt the soil community since the odds are pretty high that there isn't one. (Do not be fooled by naturally clay soils that have not been abused; these are still often quite suitable for no-dig gardening with some care.)
- **Need for designated pathways.** No-dig gardening makes it doubly important that people and animals do not traipse through the beds. This can be hard to prevent in some garden plots. If this is your problem, consider a raised bed.
- **The need for plenty of mulch.** With no-dig gardening, mulch is absolutely essential to keep the soil healthy and the weeds in check. (The good news is that mulch is strongly recommended for any style of gardening anyway.)
- **Slower soil improvement process.** Tilling in soil amendments can create perfect soil instantly. Building the soil layer by layer takes more time, although it promotes a healthier soil structure in the long run.

- **Slugs.** If you live in a wet climate, you will want to be careful with the types of mulch you use, as a stable layer of decomposing straw or grass will invite slugs to move in. This should not be an issue in dry climates.

Yes, no-dig gardening is a natural way to build good soil. However, it requires an investment.

These two tips may make the transition to no-dig gardening easier:

- **Keep your garden small.** Except for those who engage in serious canning, most gardeners can grow all the produce they need in a remarkably limited space provided they use it efficiently. Consider Mel Bartholomew's advice from *All New Square Foot Gardening*—plan on 48 square feet for every adult and 27 square feet for every child in the family, the space to be evenly divided between salad vegetables, dinner vegetables, and vegetables for preserving, giving away, or trying out for the first time. A large no-dig garden is hopelessly unmanageable for most gardeners.
- **Consider using straw as your primary source of mulch.** High-carbon mulches like wood chips and cardboard have to absorb a great deal of nitrogen from the soil to decompose, leaving less available for your plants and increasing the risk of insect pests, not to mention reducing your harvest. Straw breaks down much quicker, keeping nutrients available in the soil where they belong. (In wet climates where slugs are a problem, you may need to consider another type of mulch, such as pine straw.)

Anaerobic Conditions

Anaerobic conditions tend to favor harmful bacteria at the expense of the good guys that keep your garden in peak form. For this reason, it is important to keep your soils well aerated. For more information on the specifics of soil aeration, please refer back to Chapter 5.

Chapter 10
Maintaining Good Soil

Once you have a healthy, vibrant garden soil, you can start to reap the benefits of your efforts. Not only will gardening become easier and more enjoyable, you will harvest produce with a taste and texture far superior to anything you have ever eaten before.

Of course, if you've put in all that work to achieve good garden soil, you'll probably want to keep it that way. So how do you maintain healthy soil?

Continued Assessment

An excellent place to start is good record-keeping. One handy tool is a garden journal.

Keeping a garden journal has many benefits:

- **Permanent record.** While you can keep gardening notes on loose sheets of paper or sticky notes, the chances of finding and referring to these notes in the future are slim to none. When your notes are in one place, whether that is a binder or a designated journal, you have access to valuable information.
- **Memory aid.** Really, are you going to remember what's going on in your garden from one year to the next? If you're like most people, the answer is probably not. Write down important information. It will save you a few headaches.

- **Simplicity.** Writing in a garden journal gives you an opportunity to condense your thoughts and observations into key information that you can use.
- **Learning tool.** By noting our successes and mistakes, we have a road map to use in future years. This helps us build expertise quickly, since we are not wasting time repeating mistakes.
- **Sharpened observation skills.** Part of becoming a green thumb is observation. If you have a journal that invites you to note your observations, you might find yourself looking for new ways to fill the pages. Your powers of observation improve, and so does your understanding of your unique garden.
- **Proof of progress.** You really are developing a green thumb, and your garden journal contains proof. A review of past journals can keep you motivated and spark ideas for overcoming current challenges.
- **Gardening memories.** If you have gardened long enough, you have undoubtedly made great memories. A glance through an old journal, particularly one that contains photos, can be a nostalgic treat.

There are many things you might write in a garden journal, such as planting dates and favorite varieties. But don't forget to include notes on your soil:

- **Crop rotations.** Don't let diseases or nutrient deficiencies build up in your soil! Hang on to your map and planting records. Being able to access last year's information is a big help. Being able to access the last three years' information is better still.
- **To-dos.** Write down your gardening chores and how often they need to be done. While you're writing down what you observed today, jot down notes on what you need to do tomorrow or in a week. Staying organized is suddenly quite easy!
- **Experiments and their results.** Are you trying something new this year? Write it down, and be sure to note the results as they arise.

Not only does writing on paper cement information in our heads, but even if we do forget we have a permanent record to refer to.

- **Weeds.** Remember, weeds are indicators of what is going on in the soil. If you're experiencing an invasion, take note!
- **Pests and diseases.** Every gardener (particularly every organic gardener) has a list of "bad guys" that they count on battling every year. Improve your warfare strategy by recording the habits and preferences of the bug or fungus in question, then list ways to deter it.
- **Test results.** Every time you test your soil, whether the test is a pricey lab test or just a hands-on examination of texture and structure, note the results. You should see improvements from year to year.

The most important thing to remember about keeping a garden journal is that it should be simple. If wrestling with a bulky binder feels complicated to you, you may very well give up on your journal before the season ends. If writing a detailed essay on your garden every day feels complicated to you, you will avoid the task like the plague.

Find a journal that invites you to jot down your thoughts. Then write down only what you are interested in remembering.

Crop Rotation

Most gardeners probably already know the basic tenet of garden crop rotations:

Do not plant the same vegetable in the same place two years in row.

The reason we observe this tenet is threefold:

- To avoid building up diseases in the soil.
- To hopefully make conditions less than ideal for insect pests.
- To avoid depleting soil nutrients.

Of course, we don't want to make our crop rotations so complicated that they are no longer practical. As long as you observe the basic tenet mentioned above, you will do well. But for those of you who love planning gardens and don't mind experimenting with more sophisticated rotations, here are a few ideas to consider.

Plant Families

Two plants are generally classified as part of the same family because they share similar characteristics. It follows, then, that plants that are part of the same family probably have similar effects on the garden soil. So why not take the basic tenet one step further? Let's restate it:

Do not plant members of the same plant family in the same place two years in a row.

The most familiar examples of our slightly modified tenet are nightshades and brassicas. It is often noted that tomatoes, peppers, and potatoes (all nightshades) shouldn't follow each other in a rotation; nor should broccoli, cabbage, and kale (all brassicas).

Other plant families generally don't have the disease risk of nightshades and brassicas, but it's still a good idea to avoid repeatedly planting the same families in the same spot.

Leaf Crop, Root Crop, Fruit Crop

Another way garden plants can be (roughly) classified is by the part of the plant we eat. Do we eat the leaves (lettuce), the roots (carrots), or the fruit (tomatoes)?

It is interesting to note that leaf crops tend to use more nitrogen than the other two groups, root crops use more potassium, and fruit crops use more phosphorus. To avoid soil depletion, avoid drawing on the reserves of the same nutrient two years in a row. For example, if you have a row of carrots running along the south side of the garden one year, you probably don't want to put potatoes there the next year, even though carrots and potatoes are not in the same family.

Beneficial Plants

A third way we can approach crop rotations in the garden is to consider the role of beneficial plants. Naturally, all garden plants are beneficial from the gardener's perspective, either as food or landscaping, but some plants promote garden health and productivity, as well. For example, peas and beans fix nitrogen from the air into the soil, while marigolds exude natural insect repellents from their roots.

Of course, we don't yet understand all the different ways garden plants interact with one another and with the soil. But perhaps we can adapt what we do know into strategic crop rotation plans. For example, after a large harvest of lettuce (a leaf crop, which means it uses a lot of nitrogen), we might replace the missing soil nutrients by planting peas in that spot. Farmers use this type of thinking to plan field crop rotations; why not try it out in the garden?

Again, rotation plans that are too complicated to use are not desirable or helpful. If your garden is in overall good health, you should have plenty of leeway to make a few rotation mistakes, so keep it simple to start. As your gardening skills improve, so will your rotation strategy.

Ongoing Soil Feeding

While you hopefully won't need to address compaction or nutrient depletion again, the task of adding nutrients, replenishing organic matter, and feeding soil microbes is never finished. The good news is that it will become easier. You will want to continue mulching, composting, cover cropping, and adding other organic fertilizers to the soil regularly.

Consider keeping a garden calendar or schedule in your journal to help you keep track of these important tasks.

A Simple Soil Improvement Routine

A very easy way to improve soil regularly is to create a simple routine and tie it to your other garden chores:

1. After the first plants start germinating in the spring, apply mulch.
2. At this point, start a compost pile and keep it going all year with pulled weeds, lawn clippings, vegetable scraps, fallen leaves, etc.
3. Every time you finish harvesting a particular plant, pull the spent plants and add them to the compost pile.
4. As space in the garden is vacated, spread one trowelful of compost per square foot on top of the soil, underneath the mulch, before replanting.
5. At the end of the growing season, replenish the mulch, spread manure, or plant a winter cover crop.

Maintaining good garden soil can be easy!

Glossary

Acidic: Having a pH below 7.0 (neutral).

Actinomycete: Beneficial bacteria that somewhat resemble fungi.

Aeration: The state of being supplied with air, or the process of introducing air.

Aerobic: Operating in an oxygen-rich environment.

Albrecht test: An advanced lab test that uses acids to extract nutrients from the soil to determine total soil nutrients, base saturation, and cation exchange capacity.

Alkaline: Having a pH above 7.0 (neutral).

Amendment: Any substance added to the soil to improve its physical properties, such as texture and structure, without necessarily adding nutrients.

Anaerobic: Operating in an oxygen-deprived environment.

Anion: A molecule with a negative electrical charge.

Atmosphere: In soil, this refers to the gaseous component, occupying roughly 25% of the soil and consisting largely of carbon dioxide.

Base saturation: The percentage of soil exchange sites that are occupied by the positively charged ions Ca2+, Mg2+, K+, and Na+. See *cation exchange capacity (CEC)*.

Brix: The weight of dissolved sugar as a percentage of the weight of the entire solution.

Broadfork: A U-shaped tool with two vertical handles and a horizontal bar with tines, used to loosen and aerate soil.

Brown material: A material high in carbon to be used in a composting process.

Casting: The excrement of the earthworm.

Cation: A molecule with a positive electrical charge.

Cation exchange capacity (CEC): The measure of the total negative surface charges in the soil, directly related to the number of positively charged ions that the soil can hold on to.

Cellulose: The starch that forms the cell walls of plants.

Chelate: A ring-like compound consisting of a metal ion held in a pincher-like manner by multiple bonds with larger molecules.

Chlorosis: Loss of pigment, leading to a bleached appearance.

Clay: A very fine-textured soil consisting of particles measuring less than 0.002 mm in size.

Cold composting: A method of producing compost slowly at temperatures below 120°F.

Compaction: The process of compressing soil particles together, reducing the amount of porous space available to be occupied by air.

Compost: Decomposed organic matter used to improve the soil; as a verb, the process of decomposing organic matter for this purpose.

Conductivity: The ability of a substance, such as soil, to transmit (conduct) electrons.

Container gardening: The process of gardening in containers, such as pots, rather than in the ground.

Cover crop: A crop planted not for harvest but as a ground cover to be cut down or tilled under, thereby improving the soil.

Crop rotation: The practice of planting a different crop in a given space from year to year to avoid nutrient depletion and the buildup of pests and diseases.

Deficiency: In gardening applications, a shortfall of a given nutrient.

Diamagnetic: Being slightly repelled by a magnet.

Diatomaceous earth: A powdery substance composed of the fossilized remains of diatoms, algae that protect themselves with a silica shell; diatomaceous earth is sometimes used as a natural insecticide.

Drainage: The ability of soil to allow water to pass through.

Enzyme: A complex protein used to speed up a chemical reaction in a living organism.

Erosion: The process of wearing away by wind and water.

Friable: Loose and crumbly, characteristic of ideal soil structure.

Green material: A material high in nitrogen to be used in a composting process.

Greensand: A greenish marine sediment largely composed of sand or sandstone, used to improve soil texture and to boost potassium levels.

Hardpan: A severely compacted layer of soil that is impenetrable to plant roots.

Hay: Forage plants, primarily grasses, harvested for animal feed; hay typically contains seedheads and is therefore not suitable for mulch.

Hot composting: A method of producing compost quickly at temperatures above 120°F.

Humus: A structureless soil component derived from completely decomposed organic matter.

Inoculant: Any substance used to introduce beneficial bacteria into the soil, either as a seed coating or by direct application to the soil.

Inorganic: Any material that is not directly derived from living organisms and that cannot decompose biologically; inorganic materials can be natural (e.g., rock) or synthetic (e.g., plastic).

Ion: An atom or molecule carrying an electric charge.

Juglone: The toxin produced by the black walnut tree.

LaMotte test: See *Reams test*.

Leach: To lose water-soluble nutrients through the process of percolation.

Legume: Any species of a group of plants that bear fruit in the form of pods and have nodules on their roots to enable bacteria to fix nitrogen in the soil.

Lignin: A rigid compound found in the cell walls of woody plant tissues.

Lime: Pulverized limestone used to alter the pH of soil.

Loam: A soil texture consisting of a balanced combination of the three major particle sizes (about 40% sand, 40% silt, and 20% clay).

Luxury consumption: Excessive uptake of a given nutrient by a plant, hindering its ability to take up and use other nutrients.

Macronutrient: A nutrient used by an organism in large quantities and therefore generally recognized as being of major importance to its growth and productivity; in plants, the macronutrients are nitrogen (N), phosphorus (P), and potassium (K).

Micronutrient: A nutrient used by an organism in small quantities, but nevertheless important to its growth and productivity.

Mulch: A substance used to cover the soil to protect it from erosion, evaporation, weed growth, and extremes of temperature, among other purposes.

Mycelium: A mass of filaments, usually microscopic, formed by a fungus.

Nematode: Any one of the many species commonly known as "roundworms," some of which are beneficial and not parasitic.

No-dig gardening: A method of gardening without tillage.

NPK: An acronym for nitrogen (N), phosphorus (P), and potassium (K).

Organic: Any material that is directly derived from living organisms and that can decompose biologically; organic materials can be derived from either plants (e.g., leaves) or animals (e.g., manure).

Organic matter: The part of the soil composed of any combination of organic materials; organic matter is also added to the soil by gardeners as a fertilizer or amendment.

Paramagnetic: Being slightly attracted to a magnet.

Parent material: The underlying layer of rock that has weathered to form the mineral component of a given soil.

PC meter: A meter used to measure the paramagnetism of soil.

Peat humus: A combination of sediments and thoroughly decayed peat moss that collects at the bottoms of bogs.

Peat moss: Partially decayed sphagnum moss.

Ped: An aggregation of soil, which in turn may be formed into larger aggregates.

Percolation: The downward movement of water through soil.

pH: A measure of the concentration of hydrogen ions in a dissolved substance.

Pine straw: Pine needles collected after being shed by a pine tree in the fall.

Raised bed: A gardening bed built up above the surface of the ground.

Reams test: An advanced lab test that uses weak acids to extract nutrients from the soil to determine the amount of nutrients actually available to plants.

Refractometer: A device that measures the refraction of light through a solution, used to determine the brix of the solution.

Rhizobium: Any of a group of bacteria that live in the soil and form mutually beneficial relationships with plant roots, most notably those of legumes.

Sand: A coarse-textured soil consisting of particles ranging from about 0.05 mm to 2 mm in size.

Silt: A moderately fine-textured soil consisting of particles ranging from 0.002 to 0.05 mm in size.

Solarization: The process of sterilizing garden soil under plastic using the heat and light of the sun.

Sphagnum: A type of moss that grows only in acidic bogs.

Square foot gardening: A method of high-yield gardening for small spaces invented by Mel Bartholomew in the 1970s; square foot gardening is so called because it divides the growing space into square feet, usually with a visible grid.

Straw: The stubble of grain crops, baled after the grain has been harvested and used for mulch, animal bedding, and other non-feed purposes.

Structure: The arrangement of aggregated soil particles.

Subsoil: The layer of soil below the surface of the ground, directly derived from weathered rock and low in organic matter.

Texture: The size of the particles in a given soil.

Topsoil: The life-sustaining layer of soil on top of the ground, partly composed of weathered minerals and partly of organic matter.

Toxicity: An excess of a given nutrient or substance, producing adverse symptoms in plants.

Vermicompost: Compost made by feeding organic matter to earthworms.

Vermiculite: A mineral produced by heating mica until it expands into coarse granules, useful for balancing the moisture content of soil and potting mix.

Helpful Resources

Books

All New Square Foot Gardening: The Revolutionary Way to Grow More in Less Space, 2nd ed.

Mel Bartholomew

Minneapolis, MN: Cool Springs Press, 2013

An alternative to soil improvement. If you can find a copy, you might also be interested in reading Bartholomew's first book, *Square Foot Gardening*. The main difference between the two books is that the original method relied on improving the soil you already have, while the new method uses boxes filled with a perfect blend of purchased ingredients to instantly achieve the ideal soil texture and fertility.

Black Walnut Toxicity

Michael N. Dana and B. Rosie Lerner

West Lafayette, IN: Purdue University Cooperative Extension Service, 1994 https://www.extension.purdue.edu/extmedia/HO/HO-193.pdf

Factsheet on walnut toxicity and its implications for gardeners, including lists of sensitive and resistant plants.

The First Book of Farming

Charles L. Goodrich

Garden City, NY: Doubleday, Page & Company, 1923

http://www.gutenberg.org/ebooks/16900

An accessible look at the science of how plants grow and interact with soil. Geared toward cash crop farmers, but contains information useful to the home gardener, as well.

Cover Crops for Vegetable Growers

Charles W. Marr, Rhonda Janke, and Paul Conway

Manhattan, KS: Kansas State University, 1998

https://bookstore.ksre.ksu.edu/Item.aspx?catId=237&pubId=883

Free PDF download on improving garden soils with cover crops. Extensive information on choosing the best cover crop for your purpose and conditions.

The Complete Compost Gardening Guide

Barbara Pleasant and Deborah L. Martin

North Adams, MA: Storey Publishing, 2008

This friendly book makes composting simple. Step-by-step instructions are provided for several unique composting methods, ensuring you will find a system just right for your needs.

Equipment

Luster Leaf Rapitest Soil Test Kit

https://homesteadontherange.com/2017/02/23/luster-leaf-rapitest-soil-test-kit/

At-home test for pH and NPK. Extremely simple to use. Not precise, but quite sufficient for the average gardener's needs.

Links

"Composting Quick Start"

Homestead on the Range

https://homesteadontherange.com/composting-quick-start/

Includes a step-by-step guide, essential information, and reader-favorite posts on composting.

"Soil Quick Start"

Homestead on the Range

https://homesteadontherange.com/soil-quick-start/

Includes a step-by-step guide to soil improvement, FAQs, and more essential information.

"Cover Crop Decision Tool"

Midwest Cover Crops Council

https://mccc.msu.edu/covercroptool/

Helpful resource for those in Illinois, Indiana, Iowa, Kansas, Michigan, Minnesota, Missouri, Nebraska, North Dakota, Ohio, Ontario, South Dakota, and Wisconsin. Helps you pick the best cover crop for your situation based on your location, goals, drainage situation, and crop of choice (warm- and cool-season vegetables are among the options).

"Web Soil Survey"

Natural Resources Conservation Service

United States Department of Agriculture

https://websoilsurvey.nrcs.usda.gov/app/WebSoilSurvey.aspx

This is a little overkill for most home gardeners, but it's a lot of fun for the soil geek. Learn all about the different types of soil on your property (or in your whole neighborhood!) and their recommended uses.

You May Also Enjoy

The Family Garden Journal: A Keepsake of Daily Plans, Observations, and Harvests

Michelle Lindsey

Develop your green thumb while creating a keepsake! *The Family Garden Journal* offers an effective way to make the most of your learning experience in the garden, empowering you to become a true green thumb. Hundreds of carefully designed and tested pages will give you room to plan and observe your garden, every day of the year.

ISBN-13: 978-0997526103

See a complete list of titles at HomesteadOnTheRange.com/our-books/.

Made in the USA
Monee, IL
13 September 2026

62979551R00095